Table of Contents

Preface 3
1.0 INTRODUCTION 4
1.1 Purpose 4
1.2 Adaptive ESS 4
1.3 Responsibilities 4
Figure 1-1 Adaptive ESS Program Phases 5

2.0 FACTS REGARDING ESS 6
2.1 Abstract 6
2.2 What Required from the ESS Process 6
2.3 The ESS Process 7
2.4 Common Misconceptions About ESS 8
2.5 ESS Program Management Guidance 9
2.5.1 Where is ESS Applicable 9
2.5.2 Benefits of Environmental Stress Screening 9
2.5.3 Planning Considerations 10
2.5.4 Management Issues 11
2.5.5 Program Management Checklist 12
2.5.5.1 Demonstration and Validation 12
2.5.5.2 Engineering and Manufacturing Development 12
2.5.5.3 Production and Development 13
2.5.5.4 Re-Procurement and Depot Level Overhaul 14
2.5.6 Guidance Summary 14
2.6 Thermal Cycling Screen Development 15
2.6.1 Thermal Survey 15
2.6.2 Thermal Survey Purpose 15
2.6.3 Thermal Survey Guidelines 16
2.6.4 Cycle Characteristics 18
2.6.5 Temperature Extremes 18
2.6.6 Rate of Change of Temperature 19
2.6.7 Dwell Time at Temperature Extremes 20
2.6.8 Stabilization Time 20
2.6.9 Soak Time 21
2.6.10 Equipment Condition 21
2.6.11 Number of Cycles 22
2.6.11.1 Temperature Cycling to Contain Following 24
2.7 Vibration Screen Development 25
2.7.1 Vibration Preface 25
2.7.2 Vibration Survey 26
2.7.3 General Technique 26
2.7.4 Configuration 26
2.7.5 Item 26
2.7.6 Level 26
2.7.7 Strategy 26
2.7.8 Excitation System 27
2.7.9 Fixturing 27

2.7.10 Selection of Measurement Location 27
2.7.11 Accelerometers 27
2.7.12 Data Acquistion 28
2.7.13 Data Acquistion Equipment 28
2.7.14 Recorder Setup 28
2.7.15 Documentation 28
2.7.16 Calibration 29
2.7.17 Data Recording and Review 29
2.7.18 Data Processing 29
2.7.19 Data Analysis Equipment 30
2.5.20 Data Analysis Parameters 30
Table 2-1 Data Analysis Parameters Log 30
2.6.21 Documentation 30
2.7.22 Procedure 30
2.7.23 Simplified Technique 31
2.7.24 Force Limiting 32
2.7.24.1 Criteria for Force Limiting 32
2.7.24.2 Derivation of Force Limits 32
2.8 Importance of Effective ESS for Reliability Assurance 33
2.9 Selecting Screening Parameters Using Equations in Paragraph 2.8 34
2.10 Costs Incurred 35
2.10.1 Costs, Risks and Results Comparison of ESS at Various Levels of Assembly 36
2.11 Baseline End-Unit ESS Thermal Cycling Profile 37
2.12 Baseline End-Unit ESS Random Vibration Profile 37
2.13 Baseline Circuit Card Assembly ESS Thermal Cycling Profile 38
2.14 Baseline Circuit Card Assembly ESS Random Vibration Profile 38

3.0 FAILURE RATE DISTRIBUTION ANALYSIS 39
3.1 Purpose 39
3.2 A Method for Failure Rate Calculation 40
3.3 Task Sequence for Failure Rate Distribution Analysis to Determine The Best Thermal Cycling . 41
3.4 Indication of Required Outgoing Reliability Related Parameters 42
3.5 SigmaPlot Derived Failure Rate Equation 43
3.6 Curve Fitting to Failure Rate and Time To Failure Data Using SigmaPlot Software 44
3.7 Failure / Time Distribution Analysis 45
3.8 Guidelines for Additional Temperature Cycles in the Event of a Failure 46
3.9 Conclusion 46
3.10 Benefits of Adaptive ESS 46

4.0 HIGHLY ACCELERATED STRESS SCREENING (HASS) 47
4.1 Introduction 47
4.2 The HASS Process and Equipment 47
4.3 Defining the Screen 47
4.4 Proof of Screen 48
4.5 Data Analysis 49

5.0 GLOSSARY of TERMS 51
6.0 INFORMATION SOURCES RELATED to ESS 54
7.0 ABOUT THE AUTHOR 59

Preface

A clear understanding of Environmental Stress Screening (ESS) requires a good definition as a baseline. The following definition addresses the key aspects of ESS:

Environmental Stress Screening of a product is a process which involves the application of one or more specific types of environmental stresses for the purpose of precipitating to hard failure, latent, intermittent, or incipient defects or flaws which would cause product failure in the use environment. The stress may be applied in combination or in sequence on an accelerated basis but within product design capabilities.

ESS isolates manufacturing problems caused by poor workmanship, faulty and/or marginal parts. It also identifies design problems if the design is inherently fragile and if qualification and reliability growth tests are too benign. The most common stimuli used in ESS are Temperature Cycling and Random Vibration.

ESS is a process rather than a test in the normal accept/reject sense. Those participating in the effort, including the contractor, should never be led to believe that a "failure" is bad and would be held against them. *ESS is intended to stimulate defects*, not to simulate the operating environment, and therefore, factory "failures" are encouraged.

The root causes of ESS failures need to be found and corrected before there is a complete manufacturing process.

Initially, ESS must be applied to 100% of the units manufactured, including repaired units. By using a closed loop feedback system, one will be able to eventually determine if the screening program should be modified. A viable ESS program must be dynamic and the screening program must be actively managed, and tailored to the particular characteristics of the equipment being screened. This includes conducting a survey to determine the mechanical and thermal characteristics of the equipment and refining the screening profiles as more information becomes available and/or designs, processes, and circumstances evolve.

If the causes of all defects can be eliminated, ESS is no longer necessary, and may be stopped. An indicator of this is the lack of fallout (i.e. failures) during a properly designed ESS process. However the lack of fallout could also indicate that the ESS is not effective in finding the type of defects present in the product. In this case, a change in the applied stresses is needed.

It is the responsibility of the reliability engineer assigned to the particular project together with a test engineer to prepare the ESS Test Plan and Procedure. These documents should be reviewed and approved by the Reliability Group Leader and the Project Manager

The Adaptive ESS process is dynamic. That is the starting ESS parameters (Stress Levels, Time) are based on a similar equipment and changed depending on the effectiveness of the screening process (See Figure 1-1, Page 5). The ESS failure distribution should be analyzed using the Chance Defective Exponential Model (CDE) described in MIL-HDBK-344A with the information in Section 3.0 to obtain the best screening parameters that will give the highest outgoing reliability at minimum cost

1.0 INTRODUCTION

1.1 Purpose

Environmental Stress Screening (ESS) is not a test or series of tests with pass/fail criteria but a screening process specifically designed to disclose weak parts and workmanship defects requiring correction. It may be applied to components, subassemblies, assemblies, or equipment (as appropriate and cost-effective). The intent is to remove defects which would otherwise cause failure during later testing or field service. by moving early (Latent) failures from the field into the factory, effective ESS programs have demonstrated two important payoffs. First, ESS has helped fielded systems achieve their inherent Reliability and Maintainability (R&M), which translates directly into improved operational performance over time. ESS has significant potential return on investment during both development, production and usage

1.2 Adaptive ESS

This document describes why Adaptive ESS is needed for products and the ESS activities that should be performed during the Initial and Full Production of the Electric/Electronic Component based End Units.

An Adaptive ESS Program should be employed, as characterized by the following three phases:

- Conceptual Definition (Phase 1)
- Implementation (Phase 2)
- Effectiveness Review (Phase 3)

The interrelationship of these phases is shown in Figure 1-1

1.3 Responsibilities

It is the responsibility of the reliability engineer assigned to the particular project together with a test engineer to prepare the ESS Test Plan and Procedure. These documents should be reviewed and approved by the Reliability Group Leader and the Project Manager

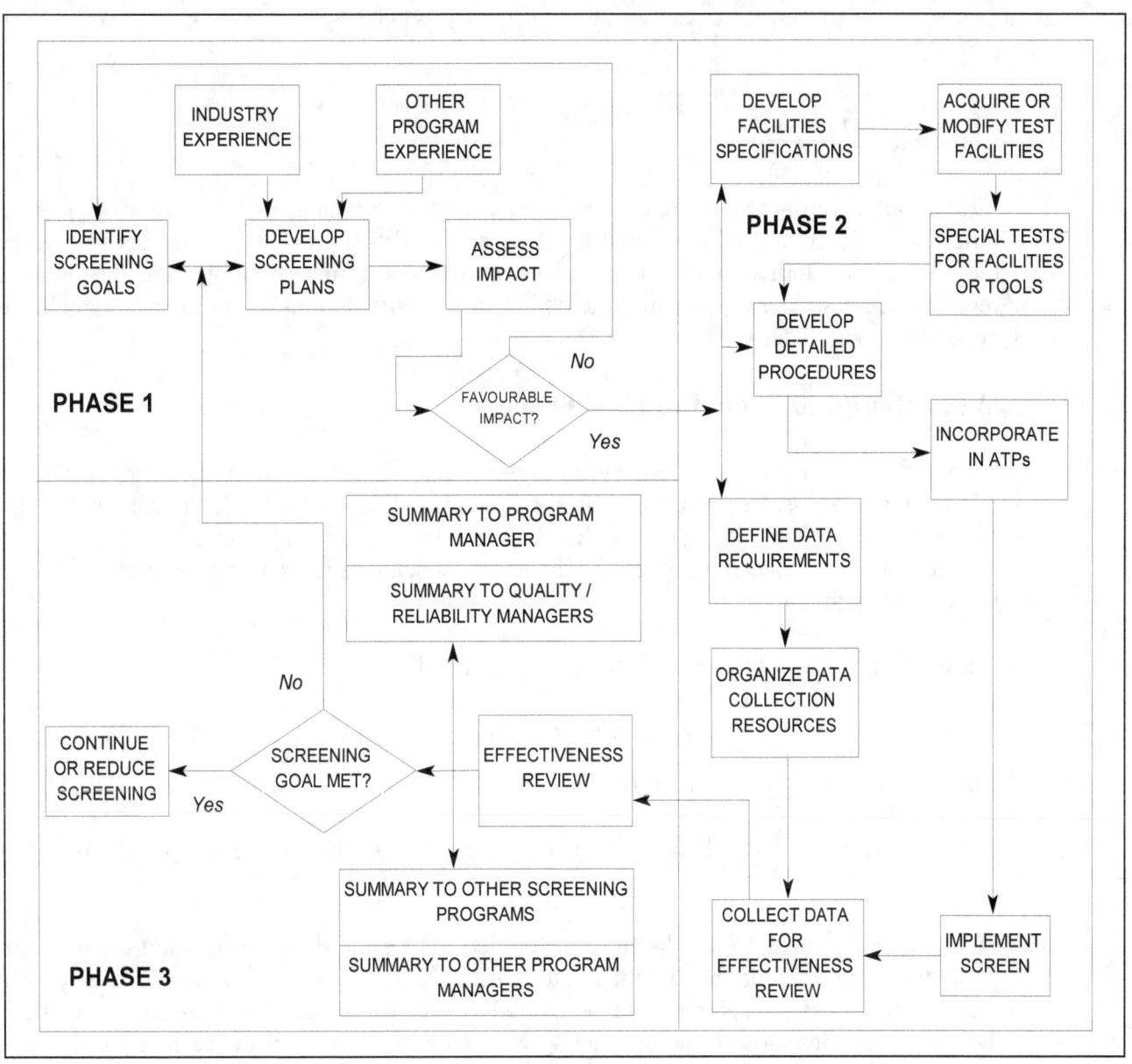

Figure 1-1 Adaptive ESS Program Phases

ATP = Acceptance Test Procedure

2.0 FACTS REGARDING ESS

2.1 Abstract

Latent Defects in a product (e.g. weak parts, faulty workmanship, design flaws) will become failures under stress. Environmental Stress Screening (ESS) prevents the defects from degrading the inherent reliability of products delivered to the customer by subjecting the product to a regimen of stress selected to cause the defects to become failures at the factory, which can then be repaired before delivery.

2.2 What Required From the ESS Process

Latent Defects in a product (e.g. weak parts, faulty workmanship, design flaws) will become failures under stress. Environmental Stress Screening (ESS) prevents the defects from degrading the inherent reliability of products delivered to the customer by subjecting the product to a regimen of stress selected to cause the defects to become failures at the factory, which can then be repaired before delivery.

The more flaws an equipment has, the higher is its failure rate.

Large Flaws ⇒ *Early Failures (Infant Mortality)*

Small Flaws ⇒ *End of Life Failures (Durability)*

ESS should remove Large Flaws without effecting much of the Small Flaw population

Stress screening tests should be fully developed during the design stage and especially before production, and should be optimized to produce cost effective benefits during production. Initially, the stress screening should be applied to 100% of all items and assembled components. This also includes those to be used as spares. However, as confidence in a particular design increases, the level of testing during design and production could be lowered; this is based upon a Statistical Test Plan. It is still important to remember that the majority of Early Life Failure (ELF)'s are caused by poor workmanship, and so screening of 100% of items prior to delivery can be regarded as essential

2.3 The ESS Process

The following flow diagram presents a typical production flow with stress screening at each stage

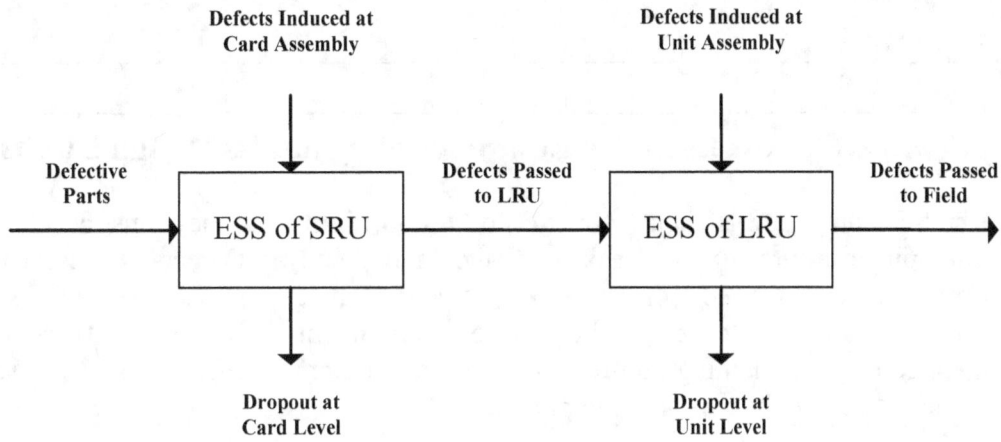

SRU = Circuit Card Assembly (CCA), Module, etc
LRU = Electronic Controller, etc

The critical importance of ESS is being able to address the types of product failure modes that may not be discovered through inspection or testing, such as:

1. Parameter drifts
2. Printed circuit board shorts and opens
3. Incorrect parts installation
4. Contaminated parts
5. Hermetic seal failures
6. Foreign material contamination
7. Cold solder joints
8. Defective parts

Operational Reliability depends on the Defects Passed to Field

2.4 Common Misconceptions About ESS

Misconception:...... **ESS Is A Test**

Truth:...... ESS Is Not a Pass/Fail Test. It is not intended to validate design, which requires a failure-free simulation process. Rather, ESS is a screening process that requires stimulation to force latent defects in products to failure that would otherwise occur in the field.

Misconception:......**ESS Stresses The Equipment Beyond Its Design Limits.**

Truth:........ The Screening Levels should not exceed Design Limits, but they must be of sufficient Strength to precipitate failures due to weak parts and Manufacturing Defects at the earliest time such that corrections are most cost effective. Effective screening requires stresses of sufficient magnitude and time duration to precipitate failures from latent defects without accumulating significant damage to the remaining non-defective structural elements. ESS should not stress the equipment such that fatigue failures are precipitated.

Misconception:.... **ESS Uncovers All Design Problems.**

Truth: ..... ESS does not uncover all Design Problems. Reliability Development/Growth Testing (RDGT) is a "Test Analyze And Fix" procedure structured to find and correct design problems. A formal RDGT is time consuming for high MTBF equipment. The objective of ESS is to accelerate early failures due to inherent and manufacturing defects such that repair is accomplished during the most cost-effective period of the program.

Misconception: ...... **ESS Is Applicable Only To New Design**

Truth: .... ESS will precipitate more design related failures On new or major modified equipment than on equipment already approved and in the inventory, but this does not restrict ESS to only new or major modified equipment. Based on the Incoming Defect Density of components, failures are precipitated by ESS regardless of whether new, modified, or remanufactured existing equipment are involved. Likewise manufacturing processes (e.g. out-of-tolerance automation) can introduce defects on any equipment, not on just new or major modified.

Reference: Environmental Stress Screening - Dimitri Kececioglu, Feng-Bin Sun

2.5 ESS Program Management Guidance

The purpose of this section is to assist program managers in understanding the issues and implementing environmental stress screening. Contracting specialists will also find this section useful in integrating and implementing ESS as a part of the acquisition strategy.

2.5.1 Where is ESS Applicable?

Best design and manufacturing practice calls for the application of environmental stress screening to:

- All material acquisitions that include electrical, electronic, electro-optical, electromechanical or electrochemical components in demonstration & validation, engineering & manufacturing development and production phases

- Re-procurements and the procurement of spare and repair parts where the cost of ESS implementation can be amortized, economically or where ESS was required in original equipment

- Depot overhaul programs where opportunities exist for substantial cost savings and overhaul/repair effectiveness

- Non-Developmental Items (NDI), such as commercial off-the-shelf (NDI-COTS) and domestic or foreign military (NDI-Military) items only to the extent ESS was implemented and documented during either current or previous production. NDI items are not to be used unless they comply with all specified requirements, including ESS

- Equipment and spares that have been specifically designated to receive ESS

ESS may be applied at any manufacturing level, from piece parts to end items. It is intended to screen defects in a manner that is not harmful to properly manufactured material. Some components, such as plasma displays, vacuum tubes, etc., by nature of their design, are not amenable to either vibration or temperature screening. Hardware proven to be too fragile may be excluded, but rationale for exclusion must be approved by the government.

2.5.2 Benefits of Environmental Stress Screening

Proper application of environmental stress screening offers several benefits:

- Reduced overall life cycle cost

- On-time deliveries

- Improved reliability after delivery

- Improved user confidence and/or satisfaction

- Reduced support costs

- Improved readiness

- Improved production process

While these benefits far outweigh the costs of implementation, they do not come without a penalty. ESS must be implemented early in the program and closely supervised throughout. It will take time and commitment of the senior managers, because the benefits are long term but the requirements for people and funds occur early in the program.

ESS is normally conducted during the manufacturing process to detect latent defects in parts and workmanship, but may also disclose design limitations that were not detected during qualification and engineering tests. In addition, there are distinct benefits to conducting ESS during development as well. A considerable percentage of the failures encountered during a reliability growth (test, analyze and fix) test program may be caused by poor workmanship and defective parts. These non-design-related failures can mask design-related failures can cause schedule slippage, and can adversely affect performance. By screening the item prior to this testing, these adverse effects can be minimized. It is virtually impossible to achieve design reliability without reducing to a minimum the reliability degradation due to screenable flaws.

2.5.3 Planning Considerations

It is imperative that ESS resources, training requirements, and detailed plans (including levels of assembly and defined profiles) are in place when production begins. Therefore, it is desirable to reach this state during engineering & manufacturing development, so that hardware for qualification and reliability growth testing is of higher quality and can be screened (to prevent failures that are not design related). This implies that experimentation and planning should begin early.

The cost of rework in manufacturing escalates by orders of magnitude as the assembly process proceeds from piece part level to printed wiring assembly/module, unit, system, and to the user. Finding defects at the lowest possible level of assembly will tend to minimize rework costs by reducing corrective action time. However, some flaw types manifest themselves only at the higher levels of assembly. Tailoring the screen by means of the vibration and thermal characteristics of the hardware coupled with defect population at each level of assembly is essential.

This document presents many management and technical details to be considered and some of the trade-off decisions that will vary with specific programs. The guidance presented may be limited in some areas but there is not intent to make this a textbook on the many facets of ESS. Because of the obvious cost, schedule and performance impacts relative to the ESS decisions, both government and contractor program managers must not treat their decisions lightly. Where the required ESS expertise is not available or will not be in time to address these issues, ESS consultants may need to be considered. Each service has at least one organization which specializes in ESS engineering and the Institute of Environmental Sciences has published many appropriate articles.

2.5.4 Management Issues

The following ESS management issues and guidance need to be considered to increase the probability of implementing a viable ESS program:

- How critical are the items proposed for ESS and what level of quality is required? Criticality would be high if a failure of the item results in high probability of loss of life or an inability to complete a mission, high life-cycle cost, or high cost of failure.

- The quantity to be procured should be considered. Where small quantities are involved and the item does not qualify as a high criticality item as given above, then it may be cost effective to use only the relatively low cost thermal cycling screens.

- The tailoring and optimization processes described in this document may result in stress levels or other ESS parameters that are less than those of the baseline. In all cases, appropriate rationale and data should be presented to justify the ESS conditions to be applied.

- The type of random vibration should be considered. Should true random vibration excitation or a low cost alternative such as quasi-random (pneumatic) vibration excitation be used for the detection of flaws?

- The contractor's proposed ESS program plan should emphasize the following:

    - Commitment to and understanding of ESS

    - Failure Reporting And Corrective Action System (FRACAS). A FRACAS should be in place and operating.

    - Span of control for ESS. If ESS is being performed by multiple subcontractors, what is their degree of implementation?

    - Planned ESS profile optimization technique. The Services recommend one of four random vibration techniques, and one of two thermal cycling techniques. Each has both advantages and disadvantages.

    - Managerial and technical approaches to ESS. The plan should include proposed methods for determining initial screening environment, applicable assembly levels, data collection, failure analysis and corrective action, and procedures or methods to be used in altering the program.

    - Non Development Items (NDI), such as commercial off-the-shelf and domestic or foreign military items, if those items have been determined to meet government requirements.

- The Government program manager should also address the following additional issues:

    - ESS profile requirements should not be specified in the RFP. In general, it is better to allow contractors to propose an ESS profile than to specify a particular profile, unless the contract is a re-procurement and the original profile holding fixtures, vibration machine and chamber capabilities are contained or referenced on the drawings and are found to be satisfactory. (Note: the original profile may have to be modified due to changes in the production process and component manufacturing variability.)

- The ESS and quality history of the contractor

2.5.5 Program Management Checklist

The following checklist should be used in the development of a management plan for implementing ESS in each phase of the acquisition process:

2.5.5.1 Demonstration & Validation

- Establish adequate ESS funding. To facilitate this, a cost/benefit analysis should be conducted to justify funding. The basis of this analysis could be the development of a historical data base on costs to implement various screens versus return on investment (cost avoidance).

- Assess the training needs of ESS personnel and develop a plan to correct any identified training and/or qualification deficiencies.

- Determine equipment availability, adequacy, capacity, etc., to perform the intended screens.

- Identify special long lead equipment requirements (e.g., fixtures, racks, etc.).

- Determine appropriate initial profile.

- Establish a FRACAS to report and analyze faults that re precipitated out during screening.

2.5.5.2 Engineering & Manufacturing Development

- Continue to tailor, refine, and evaluate the adequacy of the ESS profile, striving for an optimum screen. The absence of fault precipitation during initial production or re-procurement may be an indication of a weak screen that needs further optimization.

- Establish or continue a closed-loop FRACAS to report and analyze faults that are precipitated out during screening.

- If a test-analyze-fix (TAAF) program is being implemented, apply ESS just prior to the start of the TAAF program, while continuing to strive for an optimum screen.

- Document ESS requirements and appropriate details such as profiles, screening equipment and fixtures as part of the product Technical Data Package (TDP). The requirements shall be referenced on the appropriate part/assembly drawings or parts list. Include in the TDP the statement: "To the extent that the profiles are equipment and/or manufacturer unique, they may have to be modified due to changes in material or production processes."

- Finalize the ESS profile before the system enters into production. The following guidelines are provided to assist program managers in determining whether or not a reasonable screening profile has been developed. One or more of the following techniques may be required.

    - Verify that the more severe temperature screening profiles are used at the lower assembly levels (e.g., printed wiring assembly, module, subsystem, etc.). A good ESS program should drive outmost faults at the lower levels where faults are more easily corrected and less costly to repair.

    - Verify that the proposed screening profiles meet or exceed the Tri-Service baseline. When the profiles do not meet or exceed the baseline, verify that rationale for this deviation is acceptable.

    - Verify that the proposed screening profile is not so severe that it is damaging to the item being stressed. By reviewing failure analyses a determination can be made whether or not a failed component has been overstressed. If the results of this review indicate that the item is being overstressed, the screening profile should be adjusted until failure analyses indicate no failures are due to overstressing. In some cases a minor design change, such as additional support for a resonant component, would be a more logical and cost effective solution.

    - Verification may be made that a screening profile is adequate by seeding known faults into an item and then determining if the proposed screening profile is adequate to precipitate them to hard failure.

The profiles should not change unless the manufacturing processes are changed, the system is redesigned, parts are changed, or a different type of screen is found to be more effective.

2.5.5.3 Production and Development

- Establish or continue a closed-loop FRACAS to report and analyze faults that are precipitated out during screening.

- Establish procedures to correct/monitor any workmanship/parts problems identified during screening. Screens help to identify processes (both in-house and vendor processes) that are "out of control."

- Provide parts failures information to parts manufacturers and require continuous improvements to reduce these deficiencies.

- Establish procedures to track fielded systems and evaluate field failure information against the effectiveness of the current screens whenever possible.

- Establish criteria acceptable to the Government on when and under what conditions 100% screening should be reduced to sampling. See continuous sampling plans in MIL-STD-1235C

2.5.5.4 Re-Procurement and Depot Level Overhaul

- Derive the same benefits of ESS in re-procurement items and depot overhauled items as initial production environments. Through the frequency of failure may be lower for depot overhaul items (infant mortality/design updates are already in place through field use), poor workmanship and bad replacement parts are still a problem in the depot overhaul environment.

    Note: Numerous applications of ESS may be harmful to equipment. Depending on the particular equipment, the ESS program and the frequency of overhaul, some useful life of the equipment may be consumed.

- Require that all equipment that are re-procurements be screened if ESS was required on the original procurement contract. System level equipment should be screened at the originally developed screen or at a government approved equivalent screen. Where original screens were not developed for replacement modules, a determination based or criticality and cost should be made to determine whether or not to develop an appropriate screen.

- Establish a FRACAS whenever there is a screening effort.

2.5.6 Guidance Summary

While individual program managers have great leeway on implementing ESS, the overall direction is clear. Top management's commitment and attention is the key element in a successful ESS program. The following summarizes the ESS guidance:

- Define contractual requirements.

- Identify a general approach to satisfy these requirements.

- Perform a cost analysis considering the following:

    – Assembly level at which to apply ESS
    – Level of automation versus manual labor
    – Specific rates of thermal change versus capital investment
    – Adequacy of available in-house random vibration equipment versus cost of off-site screening or the purchase of new equipment
    – Cost considerations of active power-on versus passive power-on screening

- Consider sampling for the ESS screen based on screening data collected, but only with customer concurrence.

- Coordinate the ESS program with other activities relating to quality and reliability.

- Ensure that a FRACAS has been implemented.

2.6 Thermal Cycling Screen Development

The thermal screens in widest use today are (1) thermal cycling, (2) steady high temperature, and (3) thermal shock. The thermal cycling screen is recognized by the IES as being the most cost effective although the other two are used in some special situations.

The thermal screens described by the two methods herein, thermal survey and heritage, should be considered only as starting profiles. The effectiveness of any screen should be evaluated by engineering analysis of the equipment and the expected flaws, using factory and field failure data, and the failure history of the equipment during and subsequent to the screen, adjusting the screen parameters as the screen matures.

2.6.1 Thermal Survey

A thermal screen is characterized by:

- Cycle characteristics
    - low temperature
    - high temperature
    - rate of change of temperature
    - dwell times at temperature extremes
- Equipment condition
    - powered or unpowered
    - monitored or unmonitored
- Number of cycles
- Level of assembly at which screen is performed

With the aid of a thermal survey, Method A tailors the cycle characteristics, equipment condition and number of cycles to the hardware to be screened.

2.6.2 Thermal Survey Purpose

Developing a temperature cycling screening profile in terms of the thermal environment to which the hardware is to be subjected establishes:

- Hardware temperature history
    - temperature range
    - temperature extremes

- stabilization criterion
- soak time at temperature extremes

- Elements of the hardware to be subjected to this temperature history. It is generally not cost effective to perform a long cycle that subjects the entire mass of the item being screened to the temperature extremes. This is especially true with items (such as units, systems, and heavy modules) having high thermal inertia. Accordingly, the designer of the thermal cycling screening profile must decide what elements (such as parts, solder joints, PWA connectors) are important to be subjected to the specified hardware temperature history. This decision is based on where in the assembly the defects are expected to be precipitated by the screen. This could be, for example, in the semiconductor parts or in the PWA connectors.

- The method of heat transfer to the item being screened, such as:
  - coolant circulated through a cold-plate thermally connected to the item
  - chamber air blown over the exterior of the item
  - conditioning fluid circulated through the item

To achieve a desired hardware thermal cycle, a certain temperature history of the heat transfer medium producing the thermal cycling is required. A thermal survey evaluates the thermal response of various elements in the hardware to changes in the temperature of the heat transfer medium. The temperature history of the heat transfer medium required to produce a desired hardware thermal response may then be developed.

2.6.3 Thermal Survey Guidelines

Ideally, a thermal survey should include the following four steps. However, developing a computer simulation may not always be practical, affordable, or necessary.

1. Perform a computer simulation.

   Develop a detailed transient thermal model of the heat transfer occurring in the thermal cycling screening setup. (This is different from thermal analysis or thermal mapping, which is the measurement of the operating temperatures of the deployed equipment in actual use.) The model should be capable of predicting, as functions of the temperature history of the heat transfer medium, the temperature histories of the electronic parts, the PWAs, and other elements in the hardware targeted for removal of screenable defects. This model should simulate:

   - The dissipations of active parts in the hardware being screened (in the case of powered equipment)
   - The thermal resistances between locations in the hardware and the heat transfer medium
   - The thermal capacitances of the elements of the hardware

Use the model to perform parametric analyses of the thermal responses of the elements in the hardware being screened to changes in the temperature of the heat transfer medium. The results of these analyses will be hardware and heat transfer medium temperature histories. These analyses will:

- Identify the elements having the slowest thermal response to the heat transfer medium.

- Evaluate the temperature rate of change of the heat transfer medium required to achieve the specified hardware temperature rate of change (a function of the velocity of the heat transfer medium).

- Evaluate the dwell time of the heat transfer medium required for stabilization of the hardware temperatures (a function of the velocity and temperature of the heat transfer medium).

2. Construct a replica of the actual screening facility. The thermal survey must be performed with a setup that replicates the thermal characteristics of the actual ESS setup in the following respects:

    - Facility

    - Mounting of hardware in chamber

    - Powering (if powered during ESS)

    - Cooling (if powered and actively cooled during ESS)

3. Instrument the important locations. Monitor and record the following quantities:

    - Temperatures

        – hardware (the thermal analysis performed in the first step will aid the selection of hardware locations at which to measure temperature)

        – heat transfer medium, such as chamber air

        – coolants (if actively cooled)

    - Flow rates (velocities)

        – heat transfer medium, such as chamber air

        – coolants (if actively cooled)

    - Power dissipations (if powered)

The hardware temperature histories typically are measured with thermocouples, which are point instruments (as distinguished from infrared thermography, with which a temperature map of an area is obtained). Data are obtained only at the preselected instrumentation locations, so it is important to instrument the important locations, with the aid of the computer simulation. Thermocouples must be electrically isolated from measurement surfaces that are electrically "hot."

4. Perform the experimental thermal survey by completing the following three distinct procedures:

- The unit is soaked cold with power off until all thermocouples have stabilized at the test temperature, then power is turned on for the soak period, and then the rise to temperature at the required rate for the chamber. The chamber temperature is held at the high temperature until all thermocouples have reached the test temperature. The data is used to establish the high temperature stabilization time.

- A similar cycle is run to establish the cold temperature stabilization time.

- Several complete cycles are run to fine tune the parameters to adjust for the shortened stabilization times.

In the same way as was done analytically in the computer simulation, measure the temperature histories as functions of the screening setup parameters. Perform at least three thermal cycles to establish a thermal steady state.

The results will be experimental plots used to establish the screening parameters required to achieve the specified hardware temperature histories. The analysis in the computer simulation should minimize the amount of iteration required in the laboratory to establish the screening setup parameters.

2.6.4 Cycle Characteristics

In characterizing the thermal cycle it is important to distinguish between the temperature histories of the hardware elements and that of the chamber air. The hardware temperature histories determine the effectiveness of the screen whereas the chamber air temperature history is the controlling element. To achieve a desired hardware temperature history, a certain temperature history of the chamber air producing the thermal cycling is required.

A thermal survey evaluates the thermal response of various elements in the hardware to changes in the temperature of the chamber air. The results of the thermal survey will be experimental plots of the thermal responses, measured at critical elements of the hardware, to changes in the temperature of the chamber air. The necessary temperature range and rate of change of the chamber air can then be identified for a desired response.

2.6.5 Temperature Extremes

The temperature extremes in a thermal cycle affect the effectiveness of the screen. The temperature range (the difference between the high and low temperatures) dictates the thermal stress/strain to which the hardware is subjected in each cycle. The number of cycles to failure varies inversely with the temperature range: The wider the range, the earlier the failure. By optimizing the temperature extremes, the screening profile designer can minimize the number of cycles required to precipitate flaws. Thus, the temperature extremes also affect the cost of the screen.

The key to selecting the temperature extremes is to stress the hardware adequately to precipitate flaws without damaging good hardware. In practice, temperature ranges from a minimum of 90°C to a maximum of 180°C have been used. Minimum values are: 125°C for modules (usually -50°C to 75°C), 110°C for units (usually -40°C to 70°C) and 100°C for systems (usually -40°C to 60°C). The following key factors should be considered for the extreme value:

- Storage temperature (high and low) limits of hardware such as the materials in printed wiring assemblies

- Maximum turn-on and operating temperatures of electronic parts

2.6.6 Rate of Change of Temperature

The temperature rate of change affects the screening effectiveness in a complicated way. It also affects the duration and thus the cost of the screen.

The physical effect of the rate of change of temperature is quite complex. If a slab of material were heated and cooled uniformly, the thermal stresses and strains would be independent of the temperature rate of change.

In thermal stress screening, however the heating/cooling is non uniform because of:

- Non uniform heat transfer to the surface of the hardware

- Thermal lags between the surface and interior of the hardware

- Non uniform thermal inertia of the various portions of the hardware

Consequently, instantaneous temperature gradients can exist throughout the hardware. These temperature gradients, and the resultant thermal stress/strains, increase with increasing temperature rate of change.

Consistent with this phenomenon, industry has found that increasing the temperature rate of change increases the screening strength up to a point. For example, the situation is more complicated for solder, which creeps at temperatures encountered in thermal stress screening. Creep, which has been identified as the major cause of solder joint failure, requires time to occur. If the temperature rate of change is too high, the thermal stress screening profile may actually be excessively benign for the purpose f precipitating defective solder joints to failure. (If properly conducted, environmental stress screening to precipitate defective solder joints in a specific set of equipment should have to be performed at only one level of assembly.)

If the chamber air temperature rate of change is too high, and/or if the dwell time is too short, and/or if the chamber air is too slow, then the part temperatures will not attain the chamber air temperature extremes. The result can be an unduly benign screen. Adequate experimentation and analysis can be used to tailor chamber conditions to achieve the desired temperatures and rates

The choice of temperature rate of change depends on the nature of the hardware and the flaws expected. A high temperature rate of change is expected to be the most effective for precipitating flaws in such elements as plated-through holes, whereas a slow rate of change with long dwells at high temperature is expected to be the most effective for precipitating flaws in solder joints. In practice the temperature rate of change varies from 10°C/min to 20°C/min with the nominal values as follows:

PWA Screening   15°C/min to 20°C/min
Unit Screening   10°C/min to 20°C/min
System Screening   10°C/min to 15°C/min

The screening strength does not increase indefinitely with increasing temperature rate of change.

2.6.7 Dwell Times at Temperature Extremes

The dwell time of the chamber air temperature consists of two elements

- The time required for the part temperatures to stabilize

- The additional time required to "soak" the hardware at the temperature extremes

2.6.8 Stabilization Time

The stabilization time required for internal components to reach the ultimate chamber temperature (chamber set point) has to be determined by the thermal survey. The choice of stabilization criterion affects the duration and thus the cost of the screen.

The recommended stabilization criterion is: stabilization has occurred when the temperatures of the slowest-responding performance-related elements in the hardware being screened are within 15% of the ultimate temperatures. During the screening of unpowered equipment, the ultimate temperatures are the chamber air high and low extremes. With powered screening, the hardware temperatures may have other values, depending on the specifics of the equipment and the setup. The designer of the profile must decide which elements of the hardware being screened (excluding magnetics) are to be monitored.

Defining stabilization as the time required for the rate of change of the part temperatures to decrease to some small specified value is not recommended. Thermal analyses indicate that this criterion can result in excessively long-duration and thus expensive screens.

The stabilization time for a specific screen, using the criterion recommended depends on the hardware being screened and the screening facility. The most important factors are the thermal inertia of the assembly being screened and the chamber air speed.

2.6.9 Soak Time

The soak period serves two purposes. First, this period allows solder to creep. The time required for solder to relax is on the order of 5 minutes. Second, for screens in which the equipment is powered and monitored, the soak periods at the temperature extremes enable functional testing to be performed to detect failures which do not manifest themselves at ambient temperature. The recommended values of the soak time are as follows:

- Unmonitored equipment: 5 minutes

- Monitored equipment: long enough for functional testing to be performed or 5 minutes, whichever is longer.

2.6.10 Equipment Condition

Detection of failures induced by the environmental stresses generally requires that the equipment be powered and monitored. Testing the equipment to detect failures should be done during application of environmental stress screening, otherwise intermittent failures will go undetected. Testing only before or after stressing results in high risk of letting the intermittent flaws remain.

Thermal cycling differs from vibration in this respect: The period of a vibration cycle is a small fraction of a second and the duration of a vibration screen is on the order of 10 minutes. During a vibration screen, there is not enough time to fully test a complex system. In contrast, the period of a thermal cycle is on the order of hours and the duration of a thermal cycling screen is on the order of several hours. In a thermal screen, therefore, one can test the system at either or both temperature extremes as well as at ambient temperature.

When powered equipment is subjected to thermal cycling, the situation is complex because of the temperature rise produced by the dissipation (heat) of the electronic parts. The relation between the opening part temperatures and the chamber air temperature depends on the specific equipment and the screening parameters. In addition to the instantaneous thermal gradients occurring in screens of unpowered equipment, additional thermal gradients occur because of t he flow of heat from the dissipating parts to the surroundings.

Some factors involved in deciding whether or not to have the equipment operating are as follows:

- A powered screen is more effective in precipitating flaws than an unpowered screen. Powering produces temperature gradients in the hardware not present in unpowered equipment. The thermal stresses/strains resulting from these thermal gradients may precipitate flaws that escape in unpowered screens.

- A powered and monitored screen may detect failures that escape in an unpowered screen (intermittent failures). Failures that do not manifest themselves in testing at ambient conditions may show up in testing at high or low temperature or during vibration. An example is a broken connection in which the pieces are touching just enough to provide continuity in the absence of thermal/vibration stresses.

- A powered and monitored screen is more expensive than an unpowered screen.

- A power-off screen at the PWA level of assembly is often used as an effective screen for latent part defects. However, it should only be considered if the PWA will see a powered screen at the next higher level of assembly.

Although details will differ for any specific item to be screened, the consensus of industry experience on the basis of technical and cost trade-off considerations is as follows:

Assembly Level	Equipment Condition
Board	Unpowered
Unit	Powered monitored
System	Powered monitored

2.6.11 Number of Cycles

As do the cycle characteristics, the choice of the number of cycles impacts the effectiveness and the duration and thus the cost of the screen.

In evaluating the effect on failure of the number of cycles, it is important to distinguish between fallout at the point of screening and subsequent failures at higher levels of assembly and in the field. ESS takes life out of good an bad equipment although the decrease in the useful life of good equipment is small with a well designed screening profile. The number of failures occurring per cycle usually begins low, rapidly increases, then decreases exponentially until stabilization. When stabilization occurs, usually an optimum number of cycles has been reached.

Thermal cycling produces thermal stresses which induce alternate expansion and contraction. The stresses and strains are highest at flaws because each flaw creates a stress riser that allows the stress to precipitate a flaw (i.e., latent defect) to hard (i.e., detectable) failure. The cyclic loading causes the flaws to grow. Eventually they become so large that they cause a complete structural failure and thus an electrical failure. For example, a cracked plated through hole eventually cracks completely around and causes an open circuit.

The lifetime of the product is governed by the number of cycles, that is, the number of stress/strain reversals. The number of cycles to failure is a decreasing function of the stress/strain range per cycle, which in turn is a monotonically increasing function of the temperature range per cycle. However, a properly designed thermal screen will precipitate failures in flawed items, while not consuming a significant portion of the useful life of good items.

For solder, the physics of failures induced by thermal cycling is more complex than for materials such as aluminum and copper. The reason is that, at the temperatures encountered in electronics equipment, solder creeps. Creep has been identified as the major cause of solder joint failures. Solder creeps at a rate that increases with increasing temperature. Consequently, the number of cycles to failure of solder joints depends on other parameters s well as temperature range. The most severe thermal cycles for solder are those in which creep has sufficient time to occur. However, a screen should avoid unnecessarily inducing creep in solder joints.

Although the selection of the number of thermal cycles is critical relative to the effectiveness and cost of the screen, the procedure to do so is controversial. What is recommended here is a practical empirical approach instead of estimating the residual fault content of an item and a corresponding screening strength necessary for an acceptable product.

The number of cycles varies with product complexity, design and process maturity and whether the other thermal screen characteristics have been carefully chosen. The recommended procedure for selecting the number of cycles is:

1. Be sure that the thermal survey and analyses have been completed to identify the most appropriate values of temperature range (high and low value), product and chamber temperature rate-of-change, dwell times, and whether powered and monitored.

2. Based on the above, select the initial number of cycles for the thermal screen from the following ranges:
    - PWA   20 TO 40 cycles
    - Unit/System   12 to 20 cycles

3. Perform thermal screen as planned. Record when failures occur, types of failures, and corrective actions taken to prevent reoccurrence. Plot failures as a function of temperature cycle. When stabilization occurs in above plot, reduce the screen number of cycles to value corresponding to stabilization.

4. Continue monitoring screen results to justify any other adjustments of screening cycles, either up or down, that may be warranted.

ESS Thermal Cycling Report is a formal record of the contractor's environmental stress screening results. This report is used by the procuring activity to evaluate the effectiveness of the contractor's ESS program and monitor ESS results.

2.6.11.1 Temperature Cycling to Contain Following:

    a. Report period.
    b. Equipment nomenclature.
    c. Equipment part number.
    d. Subassembly part number (if ESS is performed at the subassembly level).
    e. Date and time of temperature cycling (at the start of each cycle).
    f. Serial number of the unit(s) subjected to temperature cycling.
    g. Elapsed time from start of temperature cycling to each failure (if applicable).
    h. Number of the cycle during which each failure occurred.
    i. Indication of point in cycle when failure occurred (hot or cold).
    j. Failed component (circuit card, module or assembly).
    k. Part number or name of failed part.
    l. Reference designation of failed part.
    m. Failure mode of failed part.
    n. Cause of failure of part.
    o. Corrective action required, taken or planned.
    p. Analysis of results to determine screening effectiveness.
    q. Any recommended changes to the ESS procedures or program.
    r. Temperature Cycling Equipment:
        1) Identification by model number and manufacturer of the temperature chamber.
        2) Maximum and minimum temperatures.
        3) Maximum and minimum rate of change of temperature.
        4) Description of procedure used to perform the temperature cycling.

2.7 Vibration Screen Development

2.7.1 Vibration Preface (*Ref: ESS Guidelines July 1993; A USA Tri-Service (Army, Navy, Air Force) Technical Brief*)

Today, true random and quasi-random vibration are used almost exclusively for ESS. True random vibration, which is well known in the ESS community, applies all frequencies in a certain bandwidth (usually 20 to 2000 Hz) and is neither cyclic nor repetitive. Quasi-random vibration, on the other hand, is a relatively new technology using pneumatically driven vibrators which generate repetitive pulses. For screening applications, several (Usually 4 to 6) of these vibrators are attached to a specially designed shaker table which is allowed to vibrate in multiple axes simultaneously. This complex motion (6 degrees of freedom vibration–3 linear axes and 3 rotational axes) is very effective in finding all types of flaws.

It is not difficult to visualize that the complex interactions possible under random vibration can induce a wider variety of relative motions in an assembly. Vibration is the area of stressing that normally precipitates latent assembly flaws caused by the undesired relative motion of parts, wires, structural elements, etc., as well as mechanical flaws that lead to propagating cracks.

Some types of flaws may be precipitated to failures by either thermal cycling or random vibration. However, it is important to note that thermal cycling and random vibration are synergistic. For example, thermal cycling following random vibration sometimes leads to detection of vibration induced failures that were not apparent immediately. There have been reported cases where a very small flaw did not propagate to the point of detectability during random vibration, but advanced to the point of detectability during subsequent thermal cycling.

The combined efforts (synergism) between vibration and thermal cycling suggests that concurrent application of the two stress types may be desirable. This combined environment is in fact sometimes used in ESS, but more often is avoided because it requires more elaborate facilities. Also, concurrent application of random vibration and thermal cycling can make it difficult to determine what caused a defect so that corrective action can be taken.

If random vibration and thermal cycling are to be conducted sequentially, random vibration would usually be done first. A more effective sequence would be five minutes of random vibration prior to thermal cycling, and another five minutes of random vibration following

There are several viable methods for developing a starting profile for vibration stress screening. Starting emphasizes that developing a screen is a dynamic process. The effectiveness of any screen should be evaluated by engineering analysis of the equipment and the expected flaws, using factory and field failure data, and the failure history of the equipment during and subsequent to the screen, adjusting the screen parameters as the screen matures.

Screens should be done in the critical axis (usually perpendicular to the plane of the printed wiring assemblies) first, with similar screens developed for the second and third axes. This procedure may eliminate vibration in the second or third axis s being ineffective in screening out defects. Where strong coupling exists between axes, all but the critical axis may be eliminated with Customer approval as not cost effective in screening out defects.

2.7.2 Vibration Survey

This is the preferred method and has been used extensively. Two techniques are available: (1) *a general* technique based on recording and analyzing the data obtained to develop the spectral responses throughout the unit being screened; and (2) a simplified techniques wherein overall g_{rms} level readings are obtained at the different sites to determine if some components are either overstressed or under stressed.

2.7.3 General Technique

The development of a random vibration stress screen is predicated on tailoring the input to achieve an acceptable response throughout the unit being screened. A vibration survey is the most logical and straightforward way to determine these responses. The spectral responses from selected accelerometer sites identify the frequencies where high responses or damping occur. The input vibration level at appropriate frequencies can then be tailored to eliminate undesired high or low responses.

2.7.4 Configuration

The vibration survey configuration should replicate the configuration for the proposed screen.

2.7.5 Item

The item must be representative of the product to be screened. It should be possible to mount accelerometers internally within the item. It should be permissible to accumulate vibration time on the item.

2.7.6 Level

The vibration survey should be conducted at an input random vibration level of 2-3gRMS (Root Means Squared). A low level sine vibration sweep can also be used to obtain a very good picture of resonance responses across the desired spectrum.

2.7.7 Strategy

The survey should be performed for each input axis or combination of input axes specified for the screen. For instance, a screen performed by the sequential excitation of three orthogonal axes requires three surveys. A screen performed as the combination of a dual axes excitation and a single axis excitation requires two surveys. A triaxial input screen requires one survey.

The End-Unit, control strategy, and the number and location of control accelerometers should be the same as for the proposed screen.

2.7.8 Excitation System

The excitation system used for the survey should be the same as for the screen.

2.7.9 Fixturing

The fixture, slip-plate, and head expander used for the survey should be the same as for the screen. It is good practice to perform a vibration survey on the mounting fixture only prior to the item survey.

2.7.10 Selection of Measurement Locations

In an exhaustive survey, vibration response would be measured at each component, wire connection, mounting screw, etc., within the item to be screened. This clearly is neither feasible nor desirable. What is desirable is to measure vibration responses at locations throughout the item that are representative of responses at a majority of the potential failure locations. Approximately 20 locations should suffice for mapping most items.

2.7.11 Accelerometers.

- Physical Characteristics

    Accelerometers should be small enough that they can be mounted in the chosen location and light enough that they do not alter the dynamic characteristics of the item. In most surveys a mix of accelerometer types can be used. The standard PWAs normally require the smallest, lightest accelerometers available so as to not alter the dynamic characteristics and to fit available mounting space.

    The acceleration in three orthogonal directions must be known for each chosen measurement location. This does not mandate a triaxial measurement at each location. A measurement from another location may be substituted for one of the triaxial measurements if the response is judged to be the same over the frequency range of interest.

- Installation

    Accelerometers should measure the input to components or parts, not the response of a particular component or part. This means placing accelerometers on the PWAs, not the components, and on the front and rear panel, not the parts mounted to the panels.

2.7.12 Data Acquisition

It is assumed that the control and response acceleration data will be recorded and played back to a spectrum analyzer for data analysis. Alternatively, if the spectrum analyzer has enough data channels, the data analysis could be performed "on-line," obviating the need to record and later play back data for spectral analysis. If a recorder is not readily available, or if the number of available accelerometer channels is limited, the survey can be accomplished in segments by analyzing the response of each available accelerometer and moving the accelerometer to another location or direction. In most cases this can be done quite efficiently with minimum impact to the overall survey.

2.7.13 Data Acquisition Equipment

The data acquisition system, i.e., accelerometers, signal conditioners, and recorder system, should have sufficient dynamic range to observe and record the response accelerations. The system should be compatible within itself and with the data analysis equipment.

2.7.14 Recorder Setup

The recorder speed should be sufficient to obtain the desired frequency response for the acquired data. For the first data acquisition run in each survey, all control accelerometers should be recorded along with the response accelerometers. For all remaining data acquisition runs in each survey, one control accelerometer should be recorded with the response accelerometers. The control accelerometer should remain the same for all remaining runs to validate repeatability in case of questionable response data.

2.7.15 Documentation

Documentation for the data acquisition should include the following information:

- Screen identification

    - program name
    - item name screening station
    - recorder
    - engineer
    - date
    - excitation system

- Channel information

    - accelerometer identification
    - accelerometer serial number
    - accelerometer sensitivity
    - charge amplifier gain
    - charge amplifier serial number

- Run information

- run identification
- frequency range and level of excitation

2.7.16 Calibration

The full scale *g* level of each channel should be estimated for each survey location prior to performing the data recording. This calculation or estimate will significantly reduce the instrumentation error caused by noise threshold or saturation.

A calibration signal, preferably a sine wave representing the full scale "*g*" level of the instrumentation, should be placed on each tape data channel. The run identification should note the voltage level, equivalent g level, and frequency of the calibration signal. The calibration should be recorded for at least two minutes after any changes in the patching of charge amplifiers to the recorder, and at any time that there is a question as to whether the input gains have been adjusted since the previous run.

It is also desirable for a broadband, approximately white noise, random signal to be recorded. The frequency range of the noise signal should extend over the frequency range of the excitation and its voltage amplitude should be within the dynamic range of the recorder. This signal, coming from one source, should be recorded simultaneously on all active data channels at the beginning of each run for a period of one minute. Record the true RMS voltage level of this signal during playback. This signal permits the frequency response of each channel and the transfer function between any two channels to be measured. Any discrepancies that are found can be compensated for during analysis.

2.7.17 Data Recording and Review

The minimum duration for recording of data should be the time necessary to calculate Acceleration Spectral Density (ASD) functions over the desired frequency range, using 50 averages. This minimum time will vary, depending on the analysis block size and bandwidth, the number of channels processed simultaneously, and the analyzer computational speed. The entire run should be recorded if the screen is a non stationary process. The data should be reviewed after the run to confirm that the amplitudes are appropriate, that the waveforms appear reasonable, and that the data segment is properly identified. The gain setting of each channel should also be verified.

2.7.18 Data Processing

The end result of the vibration survey should be a collection of ASD functions on a mass storage device available for "massaging." ASD functions should be calculated for all control and response accelerometers.

2.7.19 Data Analysis Equipment

It is recommended that the data processing be performed by playing back the recorded data to a digital Fourier spectrum analyzer. The analyzer should have the capability to calculate ASD functions, label the functions, and store the functions and labels on a mass storage device such as disk or tape. Additionally, the analyzer should be able to retrieve a stored ASD, integrate the function over selected frequency ranges to obtain gRMS (Root Means Squared) values, and print the gRMS (Root Means Squared) values.

2.7.20 Data Analysis Parameters

ASD functions should be calculated with 50 averages. An analysis bandwidth of approximately 5 Hz should be used for ASD calculation over the frequency range of 20 Hz to 2000 Hz. Alternatively, a constant percentage bandwidth analyzer may be used if the bandwidth does not exceed 1/6th octave.

Table 2-1. -- Data Analysis Parameters Log

•	PROGRAM NAME	•	NUMBER OF AVERAGES
•	UNIT NAME	•	CHARGE AMPLIFIER GAIN
•	DATE	•	RECORDER CHANNEL
•	RUN IDENTIFICATION	•	MASS STORAGE DEVICE & LOCATION NUMBER
•	FREQUENCY RESOLUTION	•	MEASUREMENT IDENTIFICATION
•	FREQUENCY RANGE		

2.7.21 Documentation

Each ASD function should be stored with a unique identifier. A data analysis log should record the run information and analysis parameters shown in Table ????3.

2.7.22 Procedure

The following vibration survey procedure assumes that data is recorded on analog tape and played back to a spectrum analyzer for ASD calculation. The procedure can be modified for use with an on-line spectral analysis system.

The procedure also assumes that the excitation system is an electrodynamic shaker. For other types of excitation systems, not all steps will be relevant.

1. Record the calibration signal on all data channels of the tape recorder.

2. Record the white noise on all data channels of the tape recorder.

3. Attach any accelerometers and cables that require special treatment (disassembly of unit, clean room facilities, obstructions when installed in the fixture, etc.) to the unit.

4. Create or retrieve input specification on the controller.

5. Mount fixture to shaker table. Torque to specified values.

6. Mount control accelerometer(s) to fixture and patch to controller and data acquisition system.

7. Perform vibration dry run(s) to fixture and patch to the controller and data acquisition system.

8. Mount unit in fixture. Torque to specified values.

9. Attach remainder of response accelerometers and cables for this data run (attach accelerometers and cables for all runs if available).

10. Patch response accelerometers for this run to data acquisition system.

11. Tap check all accelerometers to verify that they are properly patched to the input of the tape recorder and that all instrumentation functions properly.

12. Install all lids, covers, and unit cabling that will be on during screening.

13. Perform vibration run, recording all data.

14. Verify that the recorded data is valid before proceeding to the next run.

15. Repeat steps 9 through 14 for remaining groups of response accelerometers.

16. Repeat steps 4 through 15 for additional surveys, if applicable.

17. Analyze recorded data to obtain ASD functions. Label and store functions on the mass storage device for later retrieval and "massaging."

Compare vibration survey response spectra against allowable stress limit criteria applicable to the assembly under evaluation. Subsequent engineering analyses may result in appropriate hardware modifications to remove vibration screening concerns. In addition, tailoring of the input spectrum is a viable alternative for reducing response maxima to within allowable stress limits. However, because extensive tailoring can adversely affect the ability to stimulate defects throughout the entire assembly, it should be viewed as the exception, not the general rule. Where warranted, temporary stiffening or damping of the assembly should be considered to eliminate the need for tailoring.

2.7.23 Simplified Technique

The simplified vibration survey technique is a modification of the general technique. The general technique is based on recording and analyzing the data obtained to develop the spectral responses throughout the unit being screened, but there are many situations where neither the equipment nor the associated trained personnel are available to do this. For these situations the general technique can be modified so that only overall gRMS (Root Means Squared) values will determine if some components are either overstressed or under stressed due to structural resonances or damping, respectively. There is some risk that responses peculiar to random vibration may be missed.

If the mean RMS response derived from multiple locations on an assembly are within +6, -3 dB of the input excitation level, no tailoring may be required. Since overall level is only a crude indication of spectral response, if the responses for individual locations differ appreciably from the mean RMS level, a vibration response survey should be conducted at an excitation level of 6 to 10 dB below the baseline screening level (2 to 3gRMS).

2.7.24 Force Limiting

Random vibration is merely a statistical sampling of vibrations that can occur over a range of frequencies. So what is done if randomly encounter a very high response during testing? This could cause unrealistically high forces to act on the Unit

In the past, what done was to "*notch*" the input spec. where we thought we might encounter extremely high responses. That is, we looked at the frequencies where the high responses occur, and lowered the input spec. by the factor that the response is above the input. Notching is good, however if the notch is off by a few Hz, a high response could still occur.

Force limiting requires that force sensors be placed between the test table and the test item. Then during random vibration, a computer monitors the forces going into the test unit. If the computer sees that the forces are rising toward a predetermined level, it will dynamically reduce the random input so that the upper force limit is not reached. Force limiting together with input spec. notching give us a pretty good feeling that our Unit will be safe during vibration screening

2.7.24.1 Criteria for Force Limiting.

The purpose of force limiting is to reduce the response of the test item at its resonances on the shaker in order to replicate the response at the combined system resonances in the flight-mounting configuration. Force limiting is most useful for structure-like test items that exhibit distinct, lightly damped resonances on the shaker. Examples are: complete spacecraft, cantilevered structures like telescopes and antennas, lightly damped assemblies such as cold stages, fragile optical components, and equipment with pronounced fundamental modes such as a rigid structure with flexible feet. The amount of relief available from force limiting is greatest when the structural impedance (effective mass) of the test item is equal to, or greater than, that of the mounting structure. Force limiting is most beneficial when the penalties of an artificial test failure are high. Sometimes this is after an initial test failure.

2.7.24.2 Derivation of Force Limits.

As the force limiting technology matures, there may eventually be as many methods of deriving force limits as there are of deriving acceleration specifications. Force spectra have typically been developed in one-third octave bands, but other bandwidths, e.g., octave or one-tenth octave bands, may also be used. Force limiting is usually restricted to the frequency regime encompassing approximately the first three modes in each axis; which might be approximately 100 Hz for a large spacecraft, 500 Hz for an instrument, or 2000 Hz for a small component. It is important to take into account that the test item resonances on the shaker occur at considerably higher frequencies than in flight. Therefore, care must be taken not to roll off the force specification at a frequency lower than the fundamental resonance on the shaker and not to roll off the specification too steeply; i.e., it is recommended that the roll-off of the force spectrum be limited to approximately 9 dB/octave.

2.8 Importance of Effective ESS for Reliability Assurance

Latent defects in a product (e.g. weak parts, faulty workmanship, design flaws), will become failures under stress. ESS prevents the defects from degrading the inherent reliability of products delivered to the customer by subjecting the product to a regimen of stress (*Thermal Cycling & Random Vibration, Power On-Off Cycling*) designed to cause the defects to become failures at the factory, which can then be repaired before delivery.

If the causes of all defects can be eliminated, ESS is no longer necessary, and may be stopped. An indicator of this is the lack of fallout (i.e. failures) during a properly designed ESS process. However the lack of fallout could also indicate that the ESS is not effective in finding the type of defects present in the product. In this case, a change in the applied stresses is needed.

The following Screening Strength equations developed by Hughes Aircraft Company for temperature and vibration screening and modified by Rome Air Development Centre (RADC) based on the data from McDonnel Aircraft Co. and Grumman Aerospace Corporation are used to calculate Screen Strengths for this particular End-Unit

---

Screening Strength For Random Vibration

$$SS_{RV}(t_{RV}) = 1 - e^{-0.0046 G^{1.71} t_{RV}}$$

where $G = g_{rms}$, rms value of applied acceleration power spectral density over the frequency spectrum

t_{RV} = Duration of applied vibration excitation, in minutes

---

Screening Strength For Temperature Cycling

$$SS_{TC}(n_{TC}) = 1 - e^{-0.0017(T+0.6)^{0.6} (\ln(e+R))^3 n_{TC}}$$

Where T = Temperature range = $T_{max} - T_{min}$, in °C

n_{TC} = Number of temperature cycles   R = Temperature rate of change
$R = 0.5 ((T_{max} - T_{min})/t_1 + (T_{max} - T_{min})/t_2)$
t_1 = Transition time from T_{min} to T_{max}, in minutes
t_2 = Transition time from T_{max} to T_{min}, in minutes

2.9 Selecting Screening Parameters Using Equations in Paragraph 2.8

End-Unit ESS Parameters		
Characteristic	**If Now**	**Future** *(Note 1)*
Temperature Range	-40°C to +71°C	-40°C to +71°C
Temp Rate of Change	5°C / Minute	10°C / Minute
Number of Cycles	16 (1 Cycle = 2 Hours)	14 (1 Cycle = 2 Hours)
Failure Free Time	16 Hours	14 Hours *(Note 2)*
Screen Strength (Thermal)	0.9805	0.9986
Acceleration Level	2 g$_{rms}$	6 g$_{rms}$
Frequency Limits	20 - 2000 Hz	20 - 2000 Hz
Axes Stimulated	1	1
Duration of Vibration	15 Minutes	10 Minutes
Screen Strength (Vibration)	0.2021	0.6265
Screen Strength (Combined)	**0.9843**	**0.9995**

Note 1: All test requirements in the selected Test Procedure remain the same except for items related to parameters shown in this column

Note 2: This parameter should be optimized analyzing the "Failure Rate / Time" distribution of failures in Thermal Cycling using the Chance Defective Exponential (CDE) Model

2.10 Costs Incurred

The cost savings are undoubtedly one of the main arguments for stress testing, and are even greater when the testing is accelerated. Reducing the number of in-service and early life failures on a piece of equipment will:

1) Reduce field expense costs;
2) Reduce warranty claims cost;
3) Reduce the amount of scrap and rework;
4) Lower the unit costs;
5) Increase product value and confidence;
6) Improve the return on investment.

The earlier that a problem is located and resolved, the lower the cost in doing so. Screening costs are lowest at component level and progressively more costly at each higher level of assembly. For example, it can cost five times more to resolve a failure before

procurement than at the design phase, 11 times more before production, and approximately 400 times more before shipment. Hence, using stress screening to precipitate failures at the beginning of a project development cycle is far easier and cheaper to perform. It is no good refusing to stress test a design; send it into production and then realizing there is a major problem at shipment stage. The costs will be excessive, needless to say the possible consequences.

Naturally there are initial cost penalties. The capital outlay on stress testing equipment will be high, as will the additional production costs, but assuming the test procedures are adequate then this cost will be comfortably exceeded by time and money saved in the long term. Any stress screening procedure will be cost effective when the costs saved through ELF recognition and early warnings of design issues outweigh those of the costs of the screening procedures.

When planning screening tests, the production cost of scrap and rework at various stages is the main parameter in determining where in the project development cycle, and to which items, stress screening should be applied. For example, given an electronic assembly where the components were to be potted, clearly screening should occur before potting so that, on failure, repair could be made. If screening occurred after potting the assembly would have to be scrapped on failure.

2.10.1 Costs, Risks and Results Comparison of ESS at Various Levels of Assembly

Latent defects in a product (e.g. weak parts, faulty workmanship, design flaws), will become failures under stress. ESS prevents the defects from degrading the inherent reliability of products delivered to the customer by subjecting the product to a regimen of stress selected to cause the defects to become failures at the factory, which can then be repaired before delivery.

The following Table provides general guidance by comparing costs, risks and results of screening at various levels of assembly

Level of Assembly	ESS Conditions/Trade-Offs						Risks/Effects			Comments
	Power Applied [1]		I/O [2]		Monitored [3]		ESS Cost	Technical		
	Yes	No	Yes	No	Yes	No		Risk	Results	
Temperature Cycling										
CCA		X	X			X	Low	Low	Poor	Conduct pre & post ESS functional test
CCA	X		X			X	High	Lower	Better	screen prior to conformal coating
CCA	X			X	X		Highest	Lowest	Best	
Unit/Box	X			X	X		Highest	Lowest	Best	If circumstances permit ESS at only
Unit/Box	X		X			X	Lower	Higher	Good	one level of assembly, implement
Unit/Box		X	X		X		Lowest	Highest	Poor	at Unit Level

Level of Assembly	ESS Conditions/Trade-Offs						Risks/Effects			Comments
	Power Applied [1]		I/O [2]		Monitored [3]		ESS Cost	Technical		
	Yes	No	Yes	No	Yes	No		Risk	Results	
Random Vibration										
CCA	X		X		X		Highest	Low	Good	Random vibration is most effective at CCA Level if:
CCA	X			X	X		High	High	Fair	1. Surface mount technology is utilized 2. CCA has large components 3. CCA is multilayer
CCA		X		X		X	Low	Highest	Poor	4. CCA cannot be effectively screened at higher assemblies
Unit/Box	X		X		X		Highest	Low	Best	Random vibration most effective at this level of assembly. Intermittent flaws
Unit/Box	X			X	X		Low	Higher	Good	most susceptible to power-on with I/O reasonably effective.
Unit/Box		X		X		X	Lowest	Highest	Poor	Decision requires cost benifit trade-off

Reference RAC Blueprint RBPR-6
Notes:

1. Power applied - at Circuit Card Assembly (CCA) level of assembly, power on during ESS is not always cost effective
2. I/O ---- equipment fully functional, with normal inputs and outputs
3. Monitored ----- monitoring key points during screen to assure proper equipment operation

2.11 Baseline End-Unit ESS Thermal Cycling Profile

Characteristic	Level of Assembly = Unit [3]
Temperature range of product	- 40°C to +70°C
Temperature rate of change of product [1]	10°C/Min. to 20°C/Min.
Stabilization Criterion	Stabilization has occurred when the temperature of the slowest-responding element in the product being screened is within 15% of the specified high and low temperature extremes. Large magnetic parts should be avoided when determining that stabilization has occurred [1]
Soak time of product at temperature extremes after stabilization • If unmonitored • If monitored	 5 Minutes Long enough to perform functional testing
Number of cycles	12 to 20
Product Condition [2]	Powered, Monitored

Notes:

1. It is the responsibility of the designer of the screening profile to decide which elements of the hardware (parts, solder joints, PWBs, connectors, etc.) must be subjected to the extreme temperatures in the thermal cycle. The temperature histories of the various elements in the hardware being screened are determined by means of a thermal survey.

2. Power is applied during the low to high temperature excursion and remain on until the temperature has stabilized at the high temperature. Power is turned off on the high to low temperature excursion until stabilization at the low temperature. Power is also turned on and off a minimum of three times at temperature extremes on each cycle.

3. Unit guidelines apply to electronic boxes and to complex modules consisting of more than one CCA (smaller electronic module)

2.12 Baseline End-Unit ESS Random Vibration Profile

Characteristic	Level of Assembly = Unit
Overall Response Level [1]	6 g_{RMS}
Frequency [2]	20 - 2000 Hz
Axes [3] (sequentially or simultaneously)	3
Duration • Axes Sequentially • Axes Simultaneously	 10 Minutes/Axis 10 Minutes
Product Condition	Powered, Monitored

Notes:

1. If no Power Spectral Density (PSD) spectrum is available for the End-Unit from the Customer, the preferred power spectral density for 6 g_{RMS} consists of 0.04 g^2/Hz from 80 to 350 Hz with a 3 dB/Octave roll off from 80 to 20 Hz and a 3 dB/Octave roll off from 350 to 2000 Hz

2. Vibration input profile for each specific application should be determined by vibration response surveys which identify the correlation between input and structural responses. Higher frequencies are usually significantly attenuated at higher levels of assembly.

3. Single axis or two axis vibration may be acceptable if data shows minimal flaw detection in the other axes

Reference RAC Blueprint RBPR-6

2.13 Baseline Circuit Card Assembly ESS Thermal Cycling Profile

Characteristic	Level of Assembly = CCA [3]
Temperature range of product	- 50°C to +75°C
Temperature rate of change of product [1]	15°C/Min. to 20°C/Min.
Stabilization Criterion	Stabilization has occurred when the temperature of the slowest-responding element in the product being screened is within 15% of the specified high and low temperature extremes. Large magnetic parts should be avoided when determining that stabilization has occurred [1]
Soak time of product at temperature extremes after stabilization • If unmonitored • If monitored	 5 Minutes Long enough to perform functional testing
Number of cycles	20 to 40
Product Condition [2]	Unpowered/Powered

Notes:

1. It is the responsibility of the designer of the screening profile to decide which elements of the hardware (parts, solder joints, PWBs, connectors, etc.) must be subjected to the extreme temperatures in the thermal cycle. The temperature histories of the various elements in the hardware being screened are determined by means of a thermal survey.

2. Power is applied during the low to high temperature excursion and remain on until the temperature has stabilized at the high temperature. Power is turned off on the high to low temperature excursion until stabilization at the low temperature. Power is also turned on and off a minimum of three times at temperature extremes on each cycle.

3. Circuit Card Assembly (CCA) guidelines apply to individual CCAs and to modules, such as flow-through electronic modules consisting of one or two CCAs bonded to a heat exchanger

2.14 Baseline Circuit Card Assembly ESS Random Vibration Profile

Characteristic	Level of Assembly = CCA [4]
Overall Response Level [1]	6 g_{RMS}
Frequency [2]	20 - 2000 Hz
Axes [3] (sequentially or simultaneously)	3
Duration • Axes Sequentially • Axes Simultaneously	 10 Minutes/Axis 10 Minutes
Product Condition	Unpowered (Powered if purchased as an item deliverable or as a spare)

Notes:

1. If no Power Spectral Density (PSD) spectrum is available for the End-Unit from the Customer, the preferred power spectral density for 6 g_{RMS} consists of 0.04 g^2/Hz from 80 to 350 Hz with a 3 dB/Octave roll off from 80 to 20 Hz and a 3 dB/Octave roll off from 350 to 2000 Hz

2. Vibration input profile for each specific application should be determined by vibration response surveys which identify the correlation between input and structural responses. Higher frequencies are usually significantly attenuated at higher levels of assembly.

3. Single axis or two axis vibration may be acceptable if data shows minimal flaw detection in the other axes

4. When random vibration is applied at the Unit level, it may not be cost effective at the CCA level. However CCAs manufactured as end item deliverables or spares may require screening using random vibration as a stimulus. Each CCA screened separately should have its own profile determined from a vibration response survey.

Reference RAC Blueprint RBPR-6

3.0 FAILURE RATE DISTRIBUTION ANALYSIS

3.1 Purpose

The purpose of this task is to investigate all ESS failures to determine the Best ESS Parameters for Thermal Cycling

The Chance Defective Exponential (CDE) Model (See Page 42) should be the chosen prediction model for failure rate distribution analysis, as the constant failure rate portion could be extracted for Acceleration Factor calculation, the average rate of defect precipitation determined for the Best Thermal Cycling Time and Failure Free Time.

CDE equation parameters (See Page 42) are obtained using SigmaPlot computer program, which uses the Marquardt-Levenberg algorithm to fit the CDE Curve

3.2 A Method For Failure Rate Calculation

3.2.1 Average Failure Rate Estimate

For any age t, the average failure rate estimate $\frac{\Lambda}{\lambda}(t)$ at that age, for a homogeneous sample of identical units which are being subjected to a reliability test while functioning in the same application and operation environment, is given by the formula:

$$\frac{\Lambda}{\lambda}(t_i) = \frac{N_f(\Delta t)}{N_T(t_i) * \Delta t}$$

where;

$N_f(\Delta t)$ = Number of units failing in the age increment Δt or on the time period from age t_i to $t_i + \Delta t$

$N_T(t_i)$ = Number of units in test, or under observation at the beginning (by definition) of the age increment Δt or at age t_i

Δt = Age increment after t_i, during which $N_f(\Delta t)$ units fail

3.2.2 Data Gathering Format for Thermal Cycling

(a) Divide the time axis into suitable time increments, 2 hour increments are suggested, starting from time 0

(b) Plot a scatter diagram of failures as they occur during the test

(c) Count the number of failures in each time slot and calculate the failure rate according to the equation in paragraph 3.2.1

(d) Tabulate the failure rate $\frac{\Lambda}{\lambda}(t_i)$ and elapsed failure times (t_i) taking the midpoint of each time slot as the point estimate of the failure time

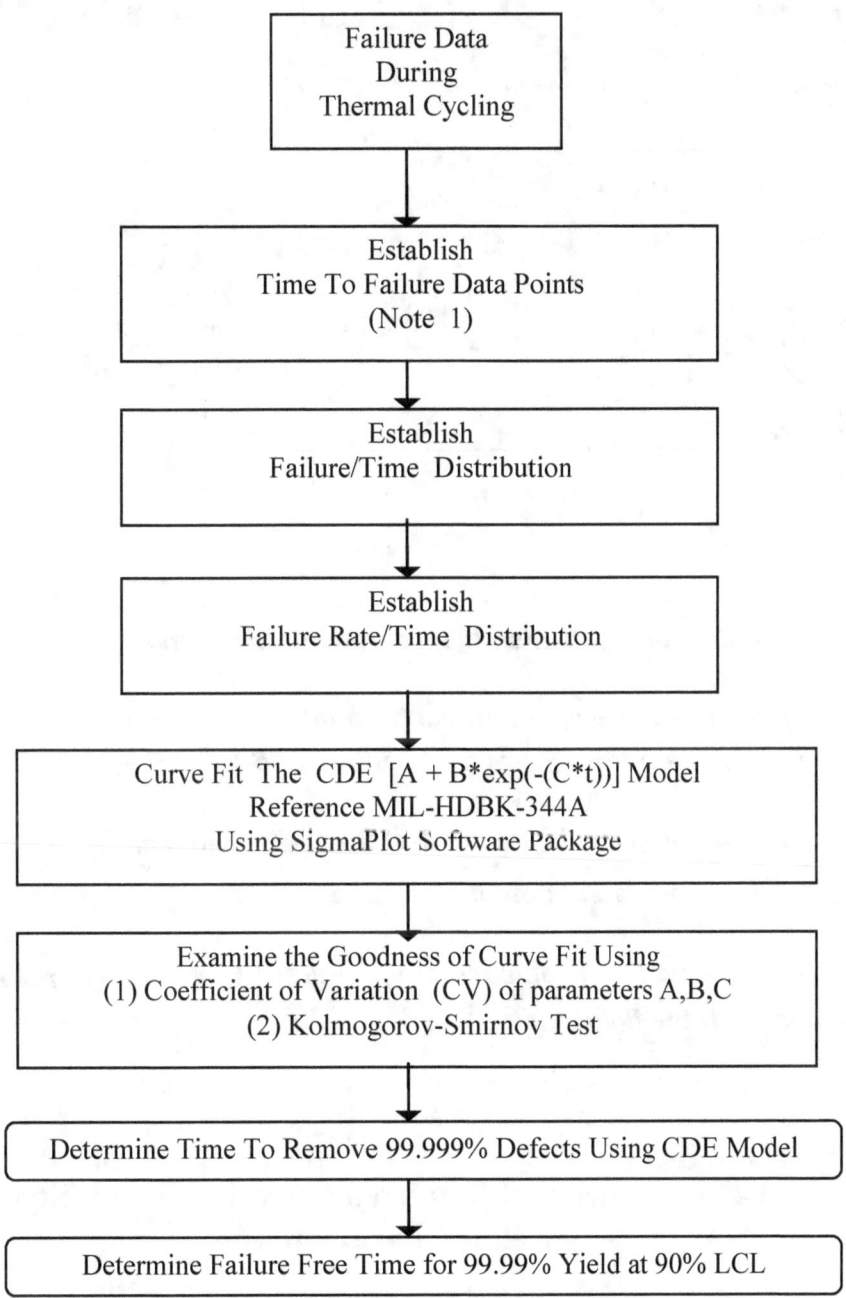

Note 1: For multiple Screen Stops in a single unit, assume Total Cumulative Thermal Energy during "No Fault Found (NFF)" and/or "Burn-in Equipment (BIE)" failure and for true failure responsible for failure precipitation

3.3 Task Sequence for Failure Rate Distribution Analysis to Determine The Best Thermal Cycling

OUTGOING RELIABILITY ASSURANCE

C = Average rate of defect precipitation under a given set of stress conditions

B/C = Incoming Defect Density (Din)

A / Normal Failure Rate = Acceleration Factor (AF)

Screening Strength (SS) = $1 - e^{-Ct}$

*Test Strength (TS) = SS * Detection Efficiency (DE)*

*Outgoing Defect Density (Dout) = Din * (1 - TS)*

Time To Remove 99.999% Defects = $(-1/C)*\ln(0.00001)$

Failure Free Time (99.99% Yield) @ 90% LCL = 7/C

Accelerated Life Tests
Environmental Stress Screening
ESS

$$FR(t) = A + Be^{-Ct}$$

Chance Defective Exponential Model

Failure Rate (FR)

t1 (variable)   Time

- **Gather failure data and establish failure rate distribution functions**

- **Starting t1 is based on equipment parts count.** **t1 is varied depending on the effectiveness of the Stress Screen (Adaptive ESS)**

- **With proper screening levels assure that there are no manufacturing process related failures. and latent defects (i.e. ESS + FFT)**   FFT = Failure Free Time

- **Performance during FFT is an accurate precursor to the kind of reliability to be expected in the field**

3.4 Indication of Required Outgoing Reliability Related Parameters

3.5 SigmaPlot Derived Failure Rate Equation

P/N 11511XX-07 ESS

Period: 01 Jan 98 - 31 Dec 98

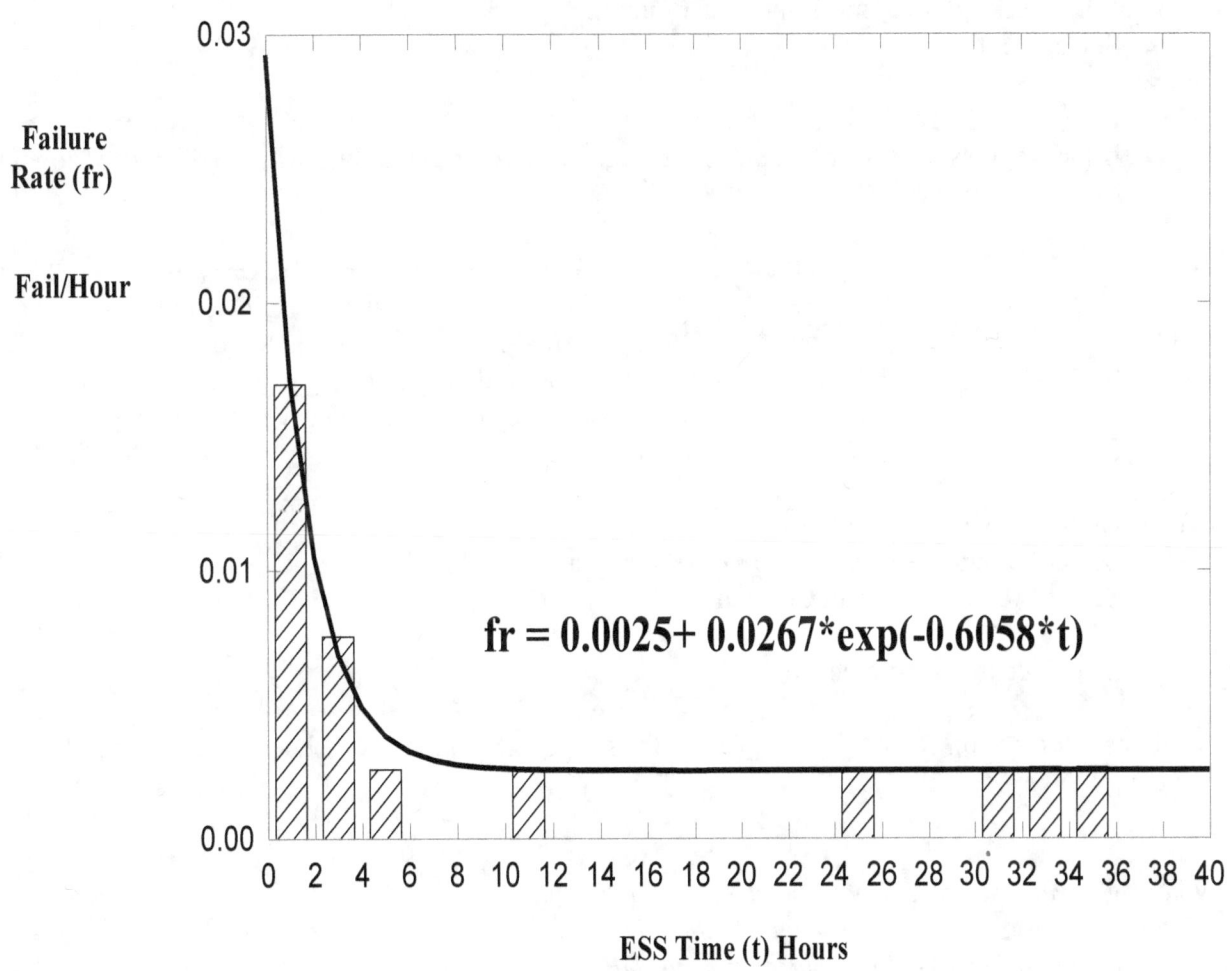

$$fr = 0.0025 + 0.0267 * \exp(-0.6058 * t)$$

CONDITION: Cumulative Thermal Energy due to previous runs And/Or NFF And/Or BIE Failure are also responsible for relevant failure precipitation

3.6 Curve ($y = y_0 + ae^{-bx}$) Fitting to Failure Rate and Time To Failure Data Using SigmaPlot Software

(1) Enter Time To Failure in Column 1, and Failure Rate in Column 2 of the Worksheet

(2) Select Statistics --------------> Select Regression Wizard
Equation Category = Exponential Decay
Equation Name = Single, 3 Parameter
Select Next
Data Format = XY Pair
Variables = x : Column 1 and y : Column 2
Select Next

(3) Regression Wizard will perform the Curve Fitting
If the Regression Wizard displays message "Converged, tolerance satisfied", write down the following numerics (tbd)

	Value	CV(%)
y_0	tbd	tbd
a	tbd	tbd
b	tbd	tbd

Select Next

> Coefficient of Variation (CV) is the normalized version of Standard Error, and is used as a gauge of the accuracy of the fitted curve parameters

(4) Go to the Results Window and
Select Parameters (i.e. y_0, a, b) to be in Column 3
Select Predicted to be in Column 4

Select Next -----> Select Next

Nonlinear Regression Details will appear on the screen. Go to the Worksheet and save the file by selecting "Save As ..."

Note: y_0 represents "A" parameter in the CDE Model
a represents "B" parameter in the CDE Model
b represents "C" parameter in the CDE Model

3.7 Failure / Time Distribution Analysis

16 failures (Component & Manufacturing related) are taken into account and a Failure Rate curve fitted (See Paragraph 3.5). The Distribution Parameters are as follows:

Goodness of Curve Fit

* Coefficient of Variation (CV) is the normalized version of Standard Error, and is used as a gauge of the accuracy of the fitted curve parameters. For this data, the CV parameters of the constants in the curve are presented below indicating a small variability of the curve fit

    Coefficient of Variation (CV) for A= 0.0025 is 12%
    Coefficient of Variation (CV) for B = 0.0267 is 9%
    Coefficient of Variation (CV) for C = 0.6058 is 12%

* A Kolmogorov-Smirnov Test is at the 20% Significance Level was conducted resulting in a Critical Value = 0.3580, Largest Difference = 0.0841. Hence the Curve adequately fits the data at 20% Significance Level

    The Kolmogorov–Smirnov test can be modified to serve as a goodness of fit test. The goodness of fit of a statistical model describes how well it fits a set of observations. Measures of goodness of fit typically summarize the discrepancy between observed values and the values expected under the model in question. Such measures can be used in statistical hypothesis testing. *Details of Kolmogorov-Smirnov Test are available in Chapter 20 of Reliability & Life Testing Handbook Volume 1, written by Dr. Dimitri Kececioglu*

Distribution Parameters

* Average Rate of Defect Precipitation = 0.6058 Defects per Hour

* Incoming Defect Density (Observed) = 0.0870 (87000 PPM)
  Incoming Defect Density (From Model) = 0.0441 (44100 PPM)

* Acceleration Factor = 66 (Based on the Normal Failure Rate = 0.000038 f/hr ..MIL-HDBK-217E)

* Screening Strength for 32 Hours ESS = 1

* Test Strength = 0.53 (Detection Efficiency = 0.53 (Observed) = 0.9 (MIL-HDBK-344))

* Outgoing Defect Density = 0.0207 (20700 PPM)

* 3σ < Capability < 4σ

* Time To Remove 99.999% Defects = 20 (Note) Hours

* Failure Free Time (99.99% Yield) @ 90% LCL = 12 (Note) Hours

Note: Values rounded to the nearest even number

3.8 Guidelines for Additional Temperature Cycles in The Event of a Failure
(Ref. NASA-CR-128908, Sep 1972)

Percentage of Total Parts Repaired/Replaced	Number of Final Consecutive Failure Free Temperature Cycles which must be survived by the Repaired/Replaced Portion of the Hardware		
	4000 Parts or more	500-2000 Parts	1-500 Parts
0 to 0.1	1	1	1
0.1 to 1	2	1	1
1 to 5	4	2	1
5 to 10	6	4	2

3.9 Conclusion

Based on the Coefficient of Variation of 12% on the Average Rate of Defect Precipitation of 0.6058 Defects/Hour, the Time To Remove 99.999% Defects lies between 17 (C = 0.6785) Hours and 22 (C = 0.5331) Hours, which is significantly less than the current Thermal Cycling Time (32 Hours total, comprising 16, 2-Hour cycles from -40 °C to +71 °C). Given this finding, it is recommended consideration be given to reducing the Thermal Cycling Time to 22 Hours and the Failure Free Time from 16 Hours to 14 Hours for P/N 115XXX

3.10 Benefits of Adaptive ESS

- Able to produce High Confidence Reliability Numerics for Reliability Guarantees
- Minimize Warranty Failures (*move failures from field to factory*) Using Optimum Outgoing Defect Densities
- Able to apply Failure Rate/Time Distributions to Predict Reliability of New Similar Electronic Controllers
- Possibility to Determine Cost Optimized ESS Times
- Improved Overall Quality of the Process and the Product
- Reduced Loading on Screening Facilities

4.0 HIGHLY ACCELERATED STRESS SCREENING (HASS)

4.1 Introduction

After the design is ruggedized through Highly Accelerated Life Testing (HALT) and the Design Verification Testing (DVT) completed, the End-Unit is ready for production. The production process can introduce many failure modes that are not related to a faulty design, and the sustaining process can certainly introduce new design problems. HASS is intended to catch these new failure modes more quickly and more effectively than traditional ESS testing done in production. The goal in HASS is to verify that no new "weak link" has crept into the product since HALT that has shifted either the operational or destruct limits found in HALT.

An important first step to setting up HASS is the completion of HALT on the product. The HASS limits have to be set based on the operational and destruct limits found in HALT. Prior to setting up HASS, it is important that corrective action has been implemented on all HALT failures and a verification HALT has been done.

4.2 The HASS Process and Equipment

The equipment used to do HASS is similar to that used in HALT, although often a larger chamber is needed to accommodate production quantities. The fixturing can be quite different in HASS, simply to accommodate the production flow. The speed with which product can be fixtured in the chamber becomes important, as well as maximizing the number of products in the chamber. Quick release clamps are often used in lieu of nuts and bolts for securing the product.

An important part of designing a fixture for HASS is the mapping of the fixture. The goal is to insure that the vibration and thermal stresses at each point in the fixture are roughly equal (although precise uniformity is not important). Mapping the fixture involves taking accelerometer and thermocouple readings on a product in each of the fixture locations. It is important the fixture is completely loaded with product for the test, since the load will affect the vibration characteristics. Thermal inconsistencies can be corrected by changing air flow through baffling or other air distribution changes. Vibration inconsistencies can be corrected through fixturing changes, with the introduction of dampening materials or changes in clamping mechanisms.

During HASS, the stresses are applied simultaneously. Typically, the product is subjected to continuous vibration while the temperature is ramped between its limits, with short dwells at the extremes.

4.3 Defining the Screen

The levels of the stresses to be applied during the screen are based on the limits found during HALT. There are two parts to the screen. The first part is the *Precipitation Screen*. This screen stresses the product beyond the *Operational Limits* and near the *Destruct Limits* found in HALT. It is intended to precipitate failures in the product due to latent defects. Because the product is being stressed beyond its operational limit, you do not expect it to function properly, so no testing is done on the product at this point. The product should be powered, however, since applied power can be a significant stress for the product in itself when combined with the other stresses of HASS. The second part of the screen is the *Detection Screen*. During the Detection screen the product is stressed to near the operational limit found in HALT. Now, the product is being functionally tested. Any hard failures induced during the Precipitation Screen should be detected, as well as any soft failures that may be induced by the stresses.

Figure 1 provides an overview of the purpose and limits of these screens. It shows the margin discovery curves, overlaid with the Precipitation and Detection Screens. The limits on the screens are set so that they are outside of the tails of the distribution of the failure mode(s) that define the operational and destruct limits for the product. Consequently, product which has no new latent failure modes should pass the screen undamaged. Any new failure mode, however, will be exposed.

There is one key problem with setting up the limits on the screens from this data - the small sample size used in HALT means that we really have no idea what the distribution looks like on these limits or where the tails may be. Consequently, a more empirical method is used. *A baseline for the stresses is derived by guard-banding the limits found in HALT. Typically, vibration is reduced by 50% and thermal excursions are reduced by 20%. These limits can be used as a starting point for the Proof of Screen process.*

4.4 Proof of Screen

Proof of Screen (PoS) is a critical part of HASS implementation. The goal of PoS is to demonstrate that the screen will reliably find defects without inducing failures or significantly reducing the life of the product. The process of PoS is fairly straightforward. A sample of product - typically a full chamber load - is run through the proposed HASS multiple times. The sample includes some seeded failures - perhaps some "no defect found" failures from field trials. The final configuration of the screen will depend on two factors - the number of cycles through the screen necessary to precipitate the seeded failures, and the number of cycles good product is able to tolerate before exhibiting end-of-life failures.

Ideally, one or two cycles through the screen will precipitate all the seeded failures. This will yield a short, efficient screen, typically lasting less than 2 hours. If seeded failures are not precipitated until several passes through the screen, then the severity of the screen should be increased. This part of the PoS verifies that the screen will reliably find defects. Multiple repetitions of the screen will demonstrate that the screen is not taking an unacceptable amount of life out of the product.

Ideally, good product will tolerate 20 to 50 passes through the screen without exhibiting failures. If end-of-life failures are seen before 20 or more cycles are complete, the screen may need to be reduced in severity. A rough estimation can be made of the amount of life being removed from the product by the screen by simply comparing the number of cycles in the proposed production screen to the number of cycles necessary to cause end of life failures to occur. For example, if the production screen consists of 2 passes through the precipitation and detection screens, and the proof of screen showed that 20 cycles through the screen induced no end-of-life failures, then the screen is removing less that 2/20, or 10%, of the useful life of your product.

The stress levels can be adjusted, or the vibration duty cycle can be changed, to achieve the proper balance between the number of cycles necessary to bring out defects versus the amount of life being taken out of the product. If stresses are increased as a result of the PoS, the PoS must be repeated on new, unstressed samples.

In reality, in can often be difficult to seed failures sufficiently to accurately verify that the screen will find defective units. Consequently, it is typically necessary to make a conservative estimate of the number of passes through the screen that are necessary, then tune the screen after a reasonable population of product has been through it. If it is found that all failures are being precipitated in the first one or two passes through the screen, then no more than two passes should be necessary. Conversely, if running 3 passes through the screen and are seeing equal failures in each pass, you should either make the screen more aggressive or increase the number of passes through the screen.

Once the HASS process is defined and proven, it is not necessarily "set in stone". Product changes can bring acceptable changes in the limits, if they are understood. However, it is always important to base our decisions on a complete failure analysis and a thorough understanding of the impact of the change. A verification HALT is a useful tool when considering these changes.

4.5 Data Analysis

There are two sources of data that should be reviewed/analyzed with regard to continuous improvement, in-house production data, and field failure analysis data.

In-house production data to be reviewed should consist of at least the following:

1. Product failure trends – Recurring product failures that are trend failures, as opposed to random failures that are non-recurring.

2. Time to device failure – time from start of screen to when a product failure is detected.

3. Point in HASS profile that device failure occurred.

4. Product first pass yield data – percent yield of product that goes through screen without failure.

5. Product failure rate/time distribution analysis – Use the Chance Defective Exponential (CDE) Model. Shown in Paragraph 3.0

This data should be reviewed and analyzed focusing on:

1. What are the product failure trends occurring in HASS?

2. What is the average time to first product failure during HASS?

3. Is there a trend as to where in the HASS profile the product failures are occurring, or are they occurring randomly?

4. What is the first pass yield of products completing HASS?

5. What constant failure rate obtained ? Calculate acceleration factor. Use average rate of

defect precipitation to determine best thermal cycling time and failure free time

Field failure analysis data to be reviewed should consist of at least the following:

1. Wear-out failures (end of life) – failures resulting from a part or group of parts reaching their end of life.

2. Infant mortality failures – failures resulting from piece part build defects or final product workmanship defects.

This data should also be reviewed and analyzed with focus on:

1. Are the wear-out failures due to HASS being too aggressive?

2. Why are the infant mortality failures not being caught in HASS?

Field failure data should be used to assess the effectiveness of changes to the HASS process.

5.0 GLOSSARY OF TERMS

Axes of Excitation - The number of axes of vibration applied during vibration screening.

Classical ESS - ESS where screening levels are not determined and continuously modified through quantitative means such as those found in Mil-Hdbk-344.

Defect: Latent Defect - An inherent or induced weakness, not detectable by ordinary means, which will either be precipitated to early failure under environmental stress screening conditions or eventually fail in the intended use environment. **Patent Defect** - An inherent or induced weakness which can be detected by inspection, functional test, or other defined means.

Defect Density - The average number of defects per item.

Detection Efficiency - A measure of the capability of detecting a patent defect.

Fallout Analysis - The study of fallout failures for the purpose of modifying screens.

Fault Replication Test Method - A method used to generate a satisfactory initial vibration screening level by gradually increasing the stress level until previously known faults are precipitated.

Final Acceptance Test - The environmental test used to validate that customer mean time between failure or failure-free requirements have been achieved. Final acceptance test is usually conducted after ESS.

Fixture - The apparatus used to mount the electronic equipment on the vibrator/shaker machine.

Flaw Precipitation Threshold Method - A method used to generate a satisfactory initial vibration screening level by performing a vibration survey and then performing detailed computations on the global responses within the test specimen. This method is also referred to as the "Tailored Spectral Response" method.

Functional Test Program - Procedures associated with testing the functionality of electronic equipment.

Heritage Screen Method - A method used to generate satisfactory initial screening levels by studying results of screening experience on similar equipment. This method is used for both vibration and thermal screening.

Mounting Scheme - The method used to affix the equipment to the shaker/vibrator.

Overall Internal Response Level Method - A method used to generate a satisfactory initial vibration screening level by performing a vibration survey and then performing simplified versions of the computations used for the flaw precipitation threshold method.

PARETO Chart - A bar chart used to highlight the few major contributors to problems vs. the trivial many contributors. The PARETO chart is based on the PARETO principle which states that 20 percent of the problems have 80 percent of the impact.

Precipitation (of Defects) - The process of transforming a latent defect into a patent defect.

Precipitation Efficiency - A measure of the capability of a screen to precipitate latent defects to failure.

Power Spectral Density - A unit of measure for random vibration. A random vibration spectrum is usually shown graphically as power spectral density in g2/Hz on the ordinate and frequency in Hz on the abscissa.

Product Reliability Verification Test (PRVT) - A test to provide confidence that field reliability will be achieved. PRVT is a segment of the ESS program implemented primarily when ESS has been nearly eliminated through corrective actions that have reduced the incoming defect densities for parts and manufacturing.

Quantitative ESS - ESS where screening parameters are determined based on models and equations which re.late required reliability to allowable remaining defect content. Such a method is outline in Mil-Hdbk-344.

Random Vibration ESS- The excitation of equipment with continuously changing frequency and peak acceleration. The equipment is exposed to a wide frequency range.

Rescreening: Incoming Parts Rescreenlng - The process of applying environmental stress screening to microcircuits, semiconductors and discrete parts at the point of receiving them from a supplier. **Repaired Equipment Rescreening** The process of screening equipment after the equipment failed as a result of ESS and was repaired back to a functional state.

Screening Regimen (or Screening Profile) - A combination of stress screens applied to an equipment, identified in the order of application {i.e., assembly, unit and system screens).

Screening Strength - The probability that a specific screen will precipitate a latent defect to failure and detect it by test, given that a latent defect susceptible to the screen is present. It is the product of precipitation efficiency and detection efficiency.

Step Stress Method - A method used to generate a satisfactory initial vibration screening level by incrementally increasing the vibration stress level until the tolerance limit is found. The tolerance limit is then used to determine the screening level.

Temperature Cycling (or Thermal Cycling) ESS - A method of ESS where equipment is exposed to high and low temperature cycling. A temperature cycling profile consists of temperature range, temperature rate of change, temperature dwell duration, number of cycles, and equipment condition (i.e., power on or off, equipment monitored or not, etc.).

Thermal Chamber- A cabinet in which hardware is placed in order to apply thermal stress to it.

Thermal Survey - The measurement of thermal response characteristics at points of interest within an equipment when temperature extremes are applied to the equipment.

Vibration Survey - The measurement of vibration response characteristics at points of interest within an equipment when vibration excitation is applied to the equipment.

Vibrator or (Electrodynamic Shaker) - A unit which an electronic equipment is attached to for conducting vibration ESS.

Yield - The probability that an equipment will pass a screen or test without failure.

6.0 INFORMATION SOURCES RELATED to ESS

(1) This Recommended Practice (RP) IEST-RP-PR001.2 supersedes IEST-RP-PR001.1, *Management and Technical Guidelines for the ESS Process* (1999).

This RP provides an overview of the techniques and considerations needed to define a viable Environmental Stress Screening (ESS) program. This RP uses language that can be used by the military, government, and commercial communities.

This RP is intended to be an important tool that can be used by the electronics industry for conducting ESS. This document applies to a knowledgeable ESS practitioner as well as to those individuals who are new to ESS. Additionally, it provides the practitioner all of the tools required to successfully plan, develop, and implement a viable ESS program.

The process elements required to successfully implement ESS are discussed in this document, as well as those process steps necessary to keep the ESS process dynamic; i.e., successful planning, implementation of ESS, the collection of failure data, the analysis of failures, and implementation of corrective action.

It is not the intent of this document to describe engineering development or formal verification environmental tests. These subjects are valuable parts of the product development process, but because of their depth and complexity are beyond the scope of this document. Rather, this document describes the application of thermal, vibration, and electrical performance as forcing functions during the ESS process because of their nearly universal utility. However, other types of forcing functions (such as pressure cycling) or performance evaluation (X-ray and infrared scanning, for example) may be appropriate in special circumstances for particular products. This document should not be construed as discouraging the application of other relevant forcing functions when appropriate This document stresses the need to expose product to forcing functions as early in the life of the product as possible. This will assist in producing the best product possible given requisite economic considerations.

It is recognized that a product will likely be composed of several different assemblies. As these assemblies are integrated and the product becomes more and more complex, the level of environmental loading that the lower level assemblies may experience can be greatly attenuated (i.e., thermal path restrictions and vibration dampening). With this in mind, and realizing that all assemblies are prone to process anomalies, this document emphasizes the need to perform ESS at the lowest practical assembly level.

This RP discusses the screening environments of temperature cycling, random vibration, and power cycling. For each of these environments, a discussion of the steps required to successfully incorporate them into the ESS program is provided. This discussion includes facility requirements, establishing screening parameters based on product response to a forcing function, and how to keep the ESS process dynamic. Furthermore, every opportunity is given to tailor ESS parameters to account for both the inherent design capability and the customer working environment as it pertains to production of the product.

This document was prepared and updated by the IEST ESS Technical Committee Working Group to provide industry and government agencies with the latest thinking and experience to assist in the development and implementation of technically sound, cost-effective stress screening programs.

(2) MIL-HDBK-344A: This HandbooK provides techniques for planning and evaluating Environmental Stress Screening (ESS) programs. The guidance contained herein departs from other approaches to ESS in that quantitative methods are used to plan and control both the cost and effectiveness of ESS programs. Handbook procedures and methodology were developed under Rome Laboratory contractual and in-house studies. Contractual efforts were performed by the Hughes Aircraft Company of Fullerton, California. The Handbook includes the guidance contained in R&M 2000 ESS Policy Letter dated 25 Jun 86.

(3) RADC-TR-86-149, FINAL TECHNICAL REPORT: ENVIRONMENTAL STRESS SCREENING (SEP-1986)., Procedures for planning, monitoring, and controlling Environmental Stress Screening (ESS) programs were developed. The technique requires estimation of the number of part and workmanship IA defects introduced in manufacture, the effectiveness of stress screens in precipitating A draft MIL-STD, prepared as part of the study, is included in the report. The data and analysis methods used in deriving the MIL-STD data tables are discussed and illustrative examples of the use of the MIL-STD procedures are also contained in the report.

(4) RADC-TR-86-138, The objective of this Guidebook is to provide techniques for planning and evaluating Environmental Stress Screening (ESS) programs. The guidance contained herein departs from other approaches to ESS in that quantitative methods are used to plan and control both the cost and effectiveness of ESS programs.

(5) IEST-RP-PR001: MANAGEMENT AND TECHNICAL GUIDELINES FOR THE ENVIRONMENTAL STRESS SCREENING (ESS) PROCESS
This Recommended Practice (RP) provides an overview of the techniques and considerations needed to define a viable Environmental Stress Screening (ESS) program. Intended to be an important tool for use by the electronics industry, this document discusses the process elements required to successfully implement ESS and the steps necessary to keep the ESS process dynamic; i.e., successful planning, implementation of ESS, the collection of failure data, the analysis of failures, and implementation of corrective action.

(6) RL-TR-94-233, IN-HOUSE REPORT: ENVIRONMENTAL STRESS SCREENING PROCESS IMPROVEMENT STUDY (DEC-1994).
This report was jointly written by engineers from the United States (US) and the Republic of Korea (ROK) under the auspices of the US/ROK Engineer and Scientist Exchange Program. The study objective was to generate an improved assembly level environmental stress screening (ESS) process through a detailed analysis and evaluation of the many existing ESS guidebooks. A process analysis technique (PAT) was used to map the processes depicted in the guidebooks. An improved process was then defined and documented using the same PAT

(7) Effectiveness of HALT, HASS and ESS for Electronic Systems, CALCE team N. M. Li, Dr. Michael Osterman. Prof. Patrick Mc Cluskey and Dr. Diginta Das, OBJECTIVES: # Determination of effectiveness of HALT, HASS and ESS for Electronic Systems # Development of guidelines for stress screens # Provide inputs to standards update for such processes for the industry

(8) NAVMAT P-9492 : NAVY MANUFACTURING SCREENING PROGRAM
This report outlines, primarily for Navy contractors, an adapted and effective manufacturing screening program consisting of temperature cycling and random vibration. With the recognition that test facility cost has been a major obstacle to the use of random vibration, a technical report, which describes in detail a proven means to generate random vibration at low cost, is included as an appendix. Together, temperature cycling and random vibration provide a most effective means of decreasing corporate costs and increasing fleet readiness.

(9) MIL-HDBK-2164A, MILITARY HANDBOOK: ENVIRONMENTAL STRESS SCREENING PROCESS FOR ELECTRONIC EQUIPMENT (19 JUN 1996) [SUPERSEDING MIL-STD-2164]., This handbook provides guidelines for Environmental Stress Screening (ESS) of electronic equipment, including environmental operation, actions takes upon detection of defects, and screening documentation. These guidelines provide for a uniform ESS process that may be utilized for effectively disclosing manufacturing defects in electronic equipment caused by poor workmanship and faulty or marginal parts. It will also identify design problems if the design is inherently fragile or if qualification and reliability growth tests were too benign or not accomplished. The most common stimuli used in ESS are temperature cycling and random vibration. A viable ESS program must be dynamic; the screening program must be actively managed, and tailored to the particular characteristics of the equipment being screened. It should be noted that there are no universal screens applicable to all equipment.

(10) Air Force Pamphlet 800-7, "USAF R&M 2000 Process". This is the Air Force document which forms the basis for the Air Force R&M program. ESS is given visibility in a detailed appendix, with specific parameters for temperature cycling and random vibration included in a chart titled "R&M 2000 Baseline Regimen."

(11) Army Materiel Command (AMC) Regulation 702-25, "AMC Environmental Stress Screening Program" This Army Regulation is the basis for ESS requirements in the Army. This regulation contains a Baseline Regimen and requires that a FRACAS be implemented. Appendix A contains a baseline Statement of Work to be used in Invitations for Bids, Requests for Proposal, and awarded contracts.

(12) DoD 4245.7-M, 'Transition from Development to Production". This document provides an excellent overview, in the Manufacturing Screening template, of the proper way to use ESS. The Manufacturing Screening template also places strong emphasis on keeping ESS dynamic and flexible through intelligent tailoring.

(13) DoD Directive 5000.1, "Defense Acquisition". This directive describes the policies which govern defense acquisition by DoD components, the major characteristics of the three decision making support systems affecting acquisition, and the acquisition responsibilities of key officials and groups. Although the directive doesn't mention ESS, it does emphasize developing reliable systems, which is the goal of ESS.

(14) DoD Instruction 5000.2, "Defense Acquisition Management Policies and Procedures". This instruction describes the procedures to be used for translating broadly stated mission needs into stable, affordable, DoD acquisition programs. It emphasizes effective acquisition planning, improved communications with users, and aggressive risk management by both Government and industry. The reliability and maintainability section (Part 6, Section C) requires that an aggressive ESS program be developed for electronic equipment and applied to engineering development and production assets.

(15) MIL-HDBK-338-1A, "Electronic Reliability Design Handbook," Volume I. This handbook, which covers all aspects of reliability program planning and execution, has a section on assembly-level ESS. The assembly-level ESS section contains a realistic approach for determining appropriate screens based on thermal and vibration surveys. The need for tailoring, continuous re-evaluation of screen cost-effectiveness, and understanding root causes of failures are continuously emphasized.

(16) MIL-HDBK-727, "Design Guidance for Producibility" This handbook, which covers all aspects of producibility, has a section on part screening. Although this tri service ESS Guidebook does not discuss Part Screening, an ESS practitioner who has the need for a detailed discussion of Part Screening may refer to MIL-HDBK-727.

(17) MIL-HDBK-781, "Reliability Test Methods, Plans, and Environments for Engineering Development, Qualification and Production". Although this is a handbook which was developed for Reliability Testing, there is a section which describes three methods for monitoring ESS (the Computed ESS Time Interval Method, the Graphical Method and the Standard ESS Method).

(18) MIL-STD-781, "Reliability Testing for Engineering Development, Qualification, and Production" This Military Standard, although developed for Reliability Qualification Testing, Reliability Growth Testing, etc., contains a Task for implementing Environmental Stress Screening in contracts.

(19) MIL-STD-785, "Reliability Program for Systems and Equipment, Development and Production". This Military Standard provides general guidance and specific tasks for reliability programs during the development, production, and initial deployment of systems and equipment Task 301 ofMIL-STD-785 provides specifics for specifying an ESS program in contracts.

(20) NAVSO P-6071, "Best Practices". NAVSO P-6071, a companion to DoD 4245.7-M, offers a very useful executive-style summary of the important issues associated with successfully using ESS. This summary is accompanied by a unique chart that contrasts the traps and consequences of some current approaches with the potential benefits of applying the Best Practices strategies.

(21) RADC-TR-87-225, "Improved Operational Readiness through Environmental Stress Screening" This technical report, developed by Rome Laboratories, contains guidelines for the application of ESS to field inventory hardware. Methods are presented for the selection of equipment for ESS application which offer significant potential for operational readiness improvement and life cycle cost reduction.

(22) RADC-TR-90-269, "Quantitative Reliability Growth Factors for ESS". This technical report, developed by Rome Laboratories, examines the measured field reliability improvements resulting from multiple cycle ESS on avionics systems. For several systems, additional cycles of ESS were applied. Through the comparison of the serialized field reliability records for those systems with and without the additional ESS cycles, an assessment of the improvement in field MTBF resulting from ESS was made.

(23) Sacramento Air Logistics Center (SM-ALC), "Environmental Stress Screening Handbook" This document was developed to provide program managers and engineers information on "how to set up an ESS program." This handbook considers often overlooked administrative as well as technical concerns such as previous contractor experience, decision criteria for ESS applicability, cost effectiveness in the production process, and contractor development of appropriate ESS methodologies.

(24) TE000-ABT-GTP-020A, "Environmental Stress Screening Requirements and ApplicationManual for Navy Electronic Equipment" This document, developed by Naval Sea Systems Command (NAVSEA), contains the basis for the NAVSEA ESS program. It is intended for use by Navy program managers as the baseline minimum ESS requirements for Statements of Work, and by design and manufacturing engineers and depot repair facilities for implementation. It contains specific information on determining the natural frequencies and displacements of PWAs. It also contains guidance on understanding the equipment's vibration and thermal responses.

(25) Tri-Service 'Technical Brief for Test Analyse And Fix (TAAF) Implementation" This document was developed by the three services in an effort to have a unified understanding of the TAAF process. This document provides in a single, concise source document the methods most likely to result in a successful TAAF program

(26) Institute of Environmental Sciences, "Environmental Stress Screening Guidelines for Assemblies, 1990" .This document differs markedly from the 1981,1984, and 1988 IES Guidelines documents in that it is procedural and tutorial in nature. This document incorporates the results of research conducted during the 1980s on the physical processes involved in ESS. The Guidelines include program management guidance, cost-effectiveness analysis techniques, descriptions of vibration and thermal survey methodologies, and ESS tailoring principles.

(27) DI-NDTI-81587, DATA ITEM DESCRIPTION: VIBRATION SURVEY REPORT (05 OCT 2000)., This report covers the results of specified vibration surveys tests performed on equipment to determine if any resonant condition exists within the equipment, and also the magnitude of resulting acceleration forces as they relate to possible over stress of assemblies or components. The report will be used by the procuring activity to determine the readiness of the equipment for environment stress screening (ESS) and reliability tests.

(28) DI-ENVR-81663, DATA ITEM DESCRIPTION: ENVIRONMENTAL STRESS SCREENING (ESS) REPORT (11 MAY 2006)., The Environmental Stress Screening (ESS) Report is a formal record of the contractors environmental stress screening results. This report is used by the procuring activity to evaluate the effectiveness of the contractors ESS program and monitor ESS results. This Data Item Description (DID) contains the format, content, and preparation instructions for the data product resulting from the work task specified in the contract.

(29) DI-ENVR-81014, DATA ITEM DESCRIPTION: ENVIRONMENTAL STRESS SCREENING (ESS) PROCEDURES AND IMPLEMENTATION PLAN (06 JUN 1990)., The ESS Procedure and Implementation Plan describes the overall ESS plan, as developed after required studies

(30) Caruso, H., " The ESS Muddle : Physics vs. Relics," Journal of the IES , pp.21-26 ,March/April 1995.
The paper provides an historical perspective on the evolution of ESS and points out activities and events leading to the current state of chaos prevalent now in the screening community. Attempt is made to clarify some common confusions and the issue of the right use of equipment (for screening) is addressed. The author regards ESS in a "state of technical and administrative disarray" and its origin and impact is one of the focal points of this paper.

(31) Environmental Stress Screening Guidelines (A Tri-Service Technical Brief) This document is the culmination of work that began in the mid-1980s when industry, with Government encouragement, initiated the revision and improvement of existing Government ESS guidelines. The Departments of the Army, Navy, and Air Force have collaborated in its preparation. It provides guidance for implementing the ESS requirements in DoD Instruction 5000.2, encouraging consistency in interpretation among all three services

7.0 ABOUT THE AUTHOR

Hilaire Ananda Perera has 39 Years North American working experience
Design, Reliability/Maintainability/Safety Engineering
https://www.linkedin.com/in/hilaireperera/

- 2007 - Present: Reliability/Maintainability/Safety Consulting Engineer at Long Term Quality Assurance (LTQA), Markham, Ontario

- 1983 - 2007: Reliability/Maintainability/Safety Engineer at Honeywell Aerospace, Mississauga, Ontario

- 1981 - 1983: Reliability/Maintainability Engineer, Philips Electronics Ltd., Telecommunications Division, Scarborough, Ontario

- 1977 - 1980: Design/Reliability Engineer, DAF Indal Ltd., Mississauga, Ontario

- 1973 - 1976: Design Engineer, Stackpole Machinary Co., Scarborough, Ontario

\# Bachelor of Science Production Engineering (1972), University of Aston, Birmingham, England \# Professional Engineer (P.Eng) 1976 to Present – Association of Professional Engineers of Ontario \# Certified (1983 - 2007) Reliability Engineer – American Society for Quality \# Honeywell Six Sigma Plus Green Belt Certified (2001), Honeywell Design For Six Sigma Certified (2003) \# Allied-Signal Aerospace 1997 Growth Award

As a Senior Engineer applied mature engineering knowledge in planning and conducting reliability engineering and related product assurance projects. Emphasized the fact that reliability is the time-based concept of quality and reliability design is an iterative process that begins with specification of reliability goals consistent with cost and performance objectives.

Introduced new approaches to achieve high product reliability. Performed to satisfy internal and external customers. Developed awareness of reliability within the organization. Served as a technical advisor on reliability issues. Assisted management in using reliability information to make. decisions for.profitability.

SPECIALTIES: ** Use probabilistic design methods (Stress/Strength, Cumulative Damage) ** Parameter Mean & Variation calculation using Tolerance Statements ** Software (MathCad, SigmaPlot, FaultrEASE, Weibull) use for analysis ** Implemented Adaptive ESS to assure a very small Outgoing Defect Density ** Promoted RAC PRISM that use stress data and Process Grade Factors as a better than MIL-HDBK-217 ** Developed a model using Weibull Parameters to.calculate.Optimum.ESS.Thermal.Cycling.Time.with.Cost

PUBLISHED PAPERS: ** Outgoing Reliability Assurance Using Chance Defective Exponential Model, AMMJ Mar 2015 ** An Introduction of Statistical Confidence Levels, AMMJ Sep 2014 ** Product Assurance Capability Quantified, Reliability Analysis Center (2Q2004) ** Adaptive ESS, Allied Signal Aerospace Company Technical Exchange Conference 25-26SEP1991 ** Optimum Cost Maintenance, Machine Design 20Jun1985 ** Reliability of Mechanical Parts, Machine Design 10Sep1987

www.ingramcontent.com/pod-product-compliance
Lightning Source LLC
Chambersburg PA
CBHW082219220526
45470CB00010B/3234